United States Nuclear Regulatory Commission

Protecting People and the Environment

NUREG-2118, Vol. 1

I0482667

Occupational Radiation Exposure at Agreement State-Licensed Materials Facilities, 1997-2010

Office of Nuclear Regulatory Research

AVAILABILITY OF REFERENCE MATERIALS
IN NRC PUBLICATIONS

United States Nuclear Regulatory Commission

Protecting People and the Environment

NUREG-2118, Vol. 1

Occupational Radiation Exposure at Agreement State-Licensed Materials Facilities, 1997-2010

Manuscript Completed: April 2012
Date Published: July 2012

Prepared by
D.E. Lewis
D.A. Hagemeyer*
Y.U. McCormick*

Office of Nuclear Regulatory Research

EDITOR'S NOTE

Staff in the U.S. Nuclear Regulatory Commission Offices of Federal and State Materials and Environmental Management Programs and Nuclear Regulatory Research assisted in the preparation of this report, serving as technical reviewers. In addition, members from the Conference of Radiation Control Program Directors, Inc. and the Organization of Agreement States provided technical comments on this report.

Comments should be directed to:

> Doris E. Lewis
> REIRS Project Manager
> Office of Nuclear Regulatory Research
> U.S. Nuclear Regulatory Commission
> Washington, DC 20555
> Phone: 301-251-7559
> E-mail Address: Doris.Lewis@nrc.gov

Paperwork Reduction Act Statement
This NUREG contains and references information collection requirements that are subject to the Paperwork Reduction Act of 1995 (44 U.S.C. 3501 et seq.). These information collections were approved by the Office of Management and Budget, approval number 3150-0014.

Public Protection Notification
The NRC may not conduct or sponsor, and a person is not required to respond to, a request for information or an information collection requirement unless the requesting document displays a currently valid OMB control number.

TABLE OF CONTENTS

LIST OF FIGURES

LIST OF TABLES

EXECUTIVE SUMMARY

On December 18, 2008, the U.S. Nuclear Regulatory Commission (NRC) staff provided a Policy Issue Notation Vote Paper, SECY-08-0197 (ADAMS Accession No. ML083360582), to the Commission which presented the regulatory and technical options of moving, or not moving, towards a greater degree of alignment of the NRC radiation protection regulatory framework with the International Commission on Radiological Protection (ICRP) Publication 103. In a Staff Requirements Memorandum dated April 2, 2009, SRM-SECY-08-0197 (ML090920103), the Commission approved the staff's recommendation to immediately begin engagement with stakeholders and interested parties to initiate development of the technical basis for possible revision of the NRC's radiation protection regulations, as appropriate and where scientifically justified. As part of the outreach to stakeholders and interested parties, NRC staff noted the need to expand the current occupational radiation dose information contained in the Radiation Exposure Information and Reporting System (REIRS) database.

Seven categories of NRC licensees are required to annually report on individual occupational exposure in accordance with Title 10 of the *Code of Federal Regulations*, Section 20.2206 (10 CFR 20.2206, "Reports of Individual Monitoring"). Specifically, these categories include commercial nuclear power reactors; industrial radiographers; fuel processors (including uranium enrichment facilities), fabricators, and reprocessors; manufacturing and distribution of byproduct material; independent spent fuel storage installations; facilities for land disposal of low-level waste; and geologic repositories for high-level waste. Because NRC has not licensed any geologic repositories for high-level waste and no NRC-licensed low-level waste disposal facilities are currently in operation, only five categories of NRC licensees report occupational radiation dose information to the REIRS database. Annually, approximately 170,000 reports on individual occupational radiation exposure are reported by NRC licensees.

The NRC's Statement on Adequacy and Compatibility of Agreement State Programs and Management Directive 5.9, "Adequacy and Compatibility of Agreement State Programs" (ML041770094) identifies the criteria that the NRC uses to determine those program elements that should be adopted by an Agreement State to maintain an adequate and compatible program. Although 10 CFR 20.2206 specifies certain categories of NRC licensees to report on individual occupational radiation exposure data, this requirement for Agreement States and their licensees is designated as a Compatibility Category D provision. Agreement States are not required to adopt Compatibility Category D provisions and reporting occupational radiation dose information is strictly voluntary on the part of the Agreement State Radiation Control Program. However, some Agreement State licensees voluntarily report their occupational radiation dose information to the NRC. Annually, approximately 3,500 reports on individual occupational radiation exposure are reported by Agreement State licensees and included in the REIRS database.

To expand the Agreement State occupational radiation dose information contained in the REIRS database, on August 6, 2010 NRC sent a letter (ML102100390) to Agreement State Radiation Control Programs. This letter requested occupational radiation dose information from industrial radiography and nuclear pharmacy licensees, from the 2000 through 2009 monitoring years. These Agreement State licensee categories were asked to submit their data to help expand the current data from NRC licensees in these two licensee categories. This information can be used by NRC staff to inform decisions regarding potential changes to the agency's current annual occupational dose limit of 5 rem to coincide with the ICRP 103 recommendation of a 2 rem annual limit.

While the NRC letter requested dose data from the 2000 through 2009 monitoring years, NRC staff recognized the need to conduct a comprehensive analysis of the Agreement State occupational radiation dose data contained in the REIRS database. This analysis would provide NRC staff with information regarding occupational dose trends among Agreement State licensees for comparison between NRC licensees within the same licensee categories. This report summarizes the Agreement State occupational radiation exposure data maintained in the REIRS database from 1997 through 2010. The information contained in this report was compiled from the annual reports submitted by nine Agreement State licensee categories. These Agreement State licensee categories are industrial radiographers; manufacturers and distributors of byproduct material; waste disposal service processing and/or repackaging; irradiators; well logging companies; sealed source facilities; measuring systems and portable gauge facilities; calibration services; and veterinary (non-human) facilities. The annual reports submitted by these licensees consist of radiation exposure records for each monitored individual. These records were analyzed for trends and are presented in this report in terms of the collective dose and the dose distribution among the monitored individuals.

From 1997 through 2010, annual reports were received from a total of 312 Agreement State licensees. Compilations of the reports submitted by the 312 Agreement State licensees indicated that 40,622 individuals were monitored, 31,704 of whom received a measurable dose (Table 3.1). The collective dose incurred by these individuals was 5,908 person-rem. The average measurable dose per individual for all Agreement State licensees calculated from the reported data was 0.19 rem. The average measurable dose is defined as the total collective dose equivalent (TEDE) divided by the number of individuals receiving a measurable dose.

ABBREVIATIONS

AEA	Atomic Energy Act
ALARA	as low as reasonably achievable
CDE	committed dose equivalent
CEDE	committed effective dose equivalent
CRCPD	Conference of Radiation Control Program Directors, Inc.
CSU	chemistry synthesis unit
DDE	deep dose equivalent
FSME	Office of Federal and State Materials and Environmental Management Programs
ICRP	International Commission on Radiological Protection
LDE	lens dose equivalent
M&D	Manufacturing and Distribution
mSv	millisievert
ND	non-detectable
NMED	Nuclear Material Events Database
NRC	U.S. Nuclear Regulatory Commission
OAS	Organization of Agreement States
REIRS	Radiation Exposure Information and Reporting System
RES	Office of Nuclear Regulatory Research
SDE-ME	shallow dose equivalent to the maximally exposed extremity
SDE-WB	shallow dose equivalent to the whole body
SI	international system of units
SNM	special nuclear material
SRM	Staff Requirements Memorandum
Sv	unit of sievert
TEDE	total effective dose equivalent
TODE	total organ dose equivalent

1 INTRODUCTION

1.1 Background

On December 18, 2008, the U.S. Nuclear Regulatory Commission (NRC) staff provided a Policy Issue Notation Vote Paper, SECY-08-0197 [Ref. 1], to the Commission which presented the regulatory and technical options of moving, or not moving, towards a greater degree of alignment of the NRC radiation protection regulatory framework with the International Commission on Radiological Protection Publication 103 [Ref. 2]. In a Staff Requirements Memorandum (SRM) dated April 2, 2009, SRM-SEC-08-0197 [Ref. 3], the Commission approved the NRC staff's recommended option to immediately begin engagement with stakeholders and interested parties. NRC staff noted the need to expand the current occupational radiation dose information contained in the Radiation Exposure Information and Reporting System (REIRS) database.

The NRC manages the REIRS database which maintains the occupational radiation dose information from specific categories of NRC licensees. Seven categories of NRC licensees are required to report annually on individual exposure in accordance with Title 10 of the *Code of Federal Regulations*, Section 20.2206 (10 CFR 20.2206, "Reports of Individual Monitoring"). Specifically, these categories include commercial nuclear power reactors; industrial radiographers; fuel processors (including uranium enrichment facilities), fabricators, and reprocessors; manufacturing and distribution of byproduct material; independent spent fuel storage installations; facilities for land disposal of low-level waste; and geologic repositories for high-level waste. Because NRC has not licensed any geologic repositories for high-level waste and no NRC-licensed low-level waste disposal facilities are currently in operation, NRC does not receive occupational radiation dose information from these two licensee categories.

To expand the Agreement State occupational radiation dose information contained in the REIRS database, NRC sent a letter (ML10210039) to Agreement State Radiation Control Programs [Ref. 4]. This letter requested occupational radiation dose information from industrial radiography and nuclear pharmacy licensees, for the 2000 through 2010 monitoring years. These Agreement State licensees were asked to submit their data to help expand the current data from NRC licensees in these two licensee categories. Currently, the REIRS database contains 100,535 records from industrial radiography and nuclear pharmacy licensees from 1997 through 2010. Nearly 39% or 38,857 are dose records from Agreement State licensees within these categories. However, twice as many industrial radiography and nuclear pharmacy licensees are licensed and regulated by Agreement States.

The purpose of this report is to examine occupational radiation exposures received under Agreement State licensees. As such, this report reflects the occupational radiation exposure data contained in the REIRS database, for 1997 through 2010, from Agreement State-licensed materials facilities.

1.2 Agreement States and NRC

Agreement States are those States that have entered into formal agreements with the NRC, pursuant to subsection 274b of the Atomic Energy Act (AEA) [Ref. 5,6], to regulate certain quantities of byproduct, source, and special nuclear material at facilities located within their borders. Agreement States issue radioactive material licenses, promulgate regulations, and enforce those regulations under the authority of each individual State's laws. Agreement States exercise their licensing and enforcement actions under direction of the governors in a manner that is adequate to protect public health and safety and compatible with the programs of the NRC. As of 2010, 36 States have signed formal agreements with the NRC [Ref. 7].

NRC assistance to States entering into agreements includes review of requests from States for 274b Agreements, or amendments to existing agreements, and meetings with States to discuss and resolve NRC review comments. NRC has the responsibility to periodically assess the Agreement State radioactive materials licensing and inspection programs through the Integrated Material Performance Evaluation Program (IMPEP) process with participation by the Organization of Agreement States (OAS) [Ref. 8]. Additionally, NRC provides early and substantive involvement of States in NRC rulemaking and other regulatory efforts; conducts training courses and workshops; evaluates technical licensing and inspection issues from Agreement States; and evaluates State rule changes.

The OAS is a nonprofit, voluntary, scientific, and professional society incorporated in the District of Columbia. The membership of OAS consists of state radiation control program directors and staff from the 36 Agreement States who are responsible for implementation of their respective Agreement State programs. The purpose of the OAS is to provide a mechanism for these Agreement States to work with each other and with the NRC on regulatory issues associated with their respective agreements.

The Conference of Radiation Control Program Directors, Inc. (CRCPD) [Ref. 9] is a 501(c)(3) non-profit, non-governmental professional organization dedicated to radiation protection. CRCPD's membership primarily consists of radiation professionals in State and local government that regulate the use of radiation sources. The purpose and goal of the CRCPD is to assist its members in their efforts to protect the public, radiation workers, and patients from unnecessary radiation exposure. CRCPD provides a forum for centralized communication on radiation protection matters between the States, Federal Government, and between individual States.

1.3 Agreement State Compatibility with NRC Regulations

The NRC's Statement on Adequacy and Compatibility of Agreement State Programs identifies the criteria that the NRC uses to determine those program elements that should be adopted by an Agreement State to maintain an adequate and compatible program. [Ref. 10] "An Agreement State radiation control program is compatible with the NRC's regulatory program when the State program does not create conflicts, duplications, gaps, or other conditions that jeopardize an orderly pattern in the regulation of agreement material (source, byproduct, and small quantities of special nuclear material as identified by Section 274b. of the Atomic Energy Act, as amended) on a nationwide basis." [Ref. 10]

An Agreement State radiation control program is adequate to protect public health and safety if administration of the program provides reasonable assurance of protection of public health and safety in regulating the use of agreement material. Accordingly, NRC program elements are established in four compatibility categories.

- Compatibility Category A
  NRC program elements in Category A are those that are basic radiation protection standards and scientific terms and definitions that are necessary to understand radiation protection concepts. The program elements adopted by an Agreement State should be essentially identical to those of NRC to provide uniformity in the regulation of agreement material on a nationwide basis.

- Compatibility Category B
  NRC program elements in Category B are those that apply to activities that have direct and significant transboundary implications. An Agreement State should adopt program elements essentially identical to those of NRC.

- Compatibility Category C
  NRC program elements in Category C are those that do not meet the criteria of Category A or B but the essential objectives of which an Agreement State should adopt to avoid conflict, duplication, gaps, or other conditions that would jeopardize an orderly pattern in the regulation of agreement material on a nationwide basis. An Agreement State should adopt the essential objectives of the NRC program elements.

- Compatibility Category D
  NRC program elements in Category D are those that do not meet any of the criteria of Category A, B, or C, above and, thus, do not need to be adopted by Agreement States for purposes of compatibility.

The requirements to report occupational radiation dose information from specific categories of licensees and participation in the NRC's REIRS database for Agreement States are designated as Compatibility Category D provisions. Because Agreement States are not required to adopt Compatibility Category D provisions, the collection and reporting of this information to the NRC is strictly voluntary on the part of the Agreement State Radiation Control Program.

2 LIMITATIONS OF THE DATA

This report summarizes the Agreement State occupational radiation exposure data maintained in the REIRS database from 1997 through 2010. Nearly seven times as many facilities are licensed and regulated by Agreement States than are licensed and regulated by NRC. There are approximately 20,000 Agreement State licensees compared with almost 3,000 NRC licensees.

From 1997 through 2009, 3,131 radiation exposure records were received by the REIRS program from Agreement State licensees. As a result of the August 6, 2010 NRC Letter, another 37,491 radiation dose records were received between 2010 and 2011. The total number of Agreement State records account for less than 1% of all records contained in the REIRS database. Occupational radiation exposure records from the following Agreement State licensee categories are in the REIRS database: industrial radiographers, manufacturers and distributors of byproduct material, waste disposal burial and service processing and/or repackaging, irradiators, well logging byproduct and/or sealed source facilities, measuring systems and portable gauge facilities, calibration services, veterinary (non-human) facilities, and other services.

The Agreement State occupational radiation dose records contained in the REIRS database can be used by NRC staff to make informed decisions regarding potential changes to the agency's current annual occupational limit of 5 rem to align with the ICRP 103 recommendation of an annual limit of 2 rem. Although the Agreement State licensee data in the REIRS database account for less than 1% of the data, an analysis can be conducted regarding the potential impact of such a regulatory change. Section 4 contains additional details regarding the potential impacts of changing the current annual occupational limit from 5 rem to 2 rem.

All of the figures compiled in this report relating to exposures and occupational doses are based on the results and interpretations of the readings of various types of radiation-monitoring devices employed by each licensee. This information, obtained from routine radiation-monitoring programs, is sufficient to characterize the radiation exposure incident to individuals' work and is used in evaluating the radiation protection program.

Monitoring requirements specified in 10 CFR 20.1502 require licensees to monitor individuals who receive or are likely to receive, in 1 year, a dose in excess of 10% of the applicable limits. For adults, the annual occupational limit is 5 rem, so 0.5 rem per year is the level above which monitoring is required. Separate dose limits have been established for minors, declared pregnant individuals, and members of the public. Monitoring is also required for any individual entering a high or very high radiation area. Depending on the administrative policy of each licensee, persons such as visitors and clerical individuals may also be provided with monitoring devices, even though the probability of their exposure to measurable levels of radiation is extremely small. It is worth noting that this report does not include compilations of non-occupational exposure, such as exposure received by medical patients from X-rays, fluoroscopy, or accelerators, which is outside NRC regulatory jurisdiction.

Many licensees elect to report the doses for every individual for whom they provided monitoring. This practice increases the number of individuals that are considered to be exposed to radiation. In an effort to account for this increase, the number of individuals reported as having "no measurable dose"[1] has been subtracted from the total number of individuals monitored in order to calculate an average dose per individual receiving a measurable dose, as well as the average dose per monitored individual.

When examining the annual statistical data, it is important to note that all of the personnel included in the report may not have been monitored throughout the entire year. Many licensees, such as industrial radiographers and nuclear pharmacies, may monitor numerous individuals for periods much less than a year. The average doses calculated from these data, therefore, are less than the average dose that an individual involved in that activity for the full year would receive. Licensees are not required to report the specific period of monitoring within the year. Many licensees report dose records for January 1 through December 31, regardless of the actual dates of monitoring. If an individual visits for only 1 day, but visits three times during the year, it is often reported as if the individual was there all year. NRC encourages reporting of the actual monitoring dates, but it is not a regulatory requirement.

All dose equivalent values in this report are given in units of rem in accordance with the general provisions for records in 10 CFR 20.2101(a). In order to convert rem into the International System of Units (SI) unit of sieverts (Sv), readers should divide the value in rem by 100. Therefore, 1 rem = 0.01 Sv. In order to convert rem into millisieverts (mSv), readers should multiply the value in rem by 10.

---

[1] The number of workers with measurable dose includes any individual with a dose greater than zero rem and does not include doses reported as "not detectable."

3 ANNUAL PERSONNEL MONITORING REPORTS

3.1 Definition of Terms and Sources of Data

Although Agreement State licensees are not required to submit exposure reports to NRC, some Agreement State licensees annually submit this data voluntarily. In addition to this voluntary submission, some Agreement State licensees also responded to the NRC Letter. The NRC Letter was a one-time request for certain categories of Agreement State licensees to submit their exposure data. As of June 2011, the NRC's REIRS database contains 40,622 occupational radiation exposure reports from Agreement State licensees. Table 3.1 summarizes the annual data that have been submitted from 1997 through 2010.

3.1.1 Number of Licensees Reporting

The number of licensees refers to the number of Agreement State licensees that filed occupational radiation exposure reports voluntarily for each year and in response to the NRC Letter. Appendix A contains additional detailed information on Agreement State licensees that have submitted information.

3.1.2 Number of Monitored Individuals

The number of monitored individuals refers to the total number of individuals that Agreement State licensees reported as being monitored for exposure to external and internal radiation during each respective year. The total number of individuals was determined from the number of unique personal identification numbers submitted per licensee.

3.1.3 Number of Individuals with Measurable Dose

The number of individuals with measurable dose includes any individual with a TEDE greater than zero rem.

3.1.4 Collective Dose

The concept of collective dose is used in this report to denote the summation of the TEDE received by all monitored individuals and is reported in units of person-rem. The collective dose is calculated by summing the TEDE for all monitored individuals. The phrase "collective dose" is used throughout this report to mean the collective TEDE, unless otherwise specified.

3.1.5 Average Measurable Dose

The average measurable dose is obtained by dividing the collective TEDE by the number of monitored individuals with measurable dose. This is the average most commonly used in this and other reports when examining trends and comparing doses received by individuals in various segments of the nuclear industry. It excludes individuals receiving zero or no detectable dose, many of whom were monitored for convenience or identification purposes.

3.2 Annual TEDE Dose Distributions

Table 3.2 provides a statistical compilation of the occupational dose reports by categories of licensees (see Section 3.4 for a description of each licensee category). The dose distributions are generated by summing the TEDE for each individual and counting the number of individuals in each dose range. In nearly every licensee category, a large number of individuals receive doses that are less than measurable, and very few doses exceed 4 rem.

Based on the 312 licensees that voluntarily submitted radiation dose data for the past 14 years, nearly 22% (8,918) of the individuals monitored received no measurable dose and 0.7% (304) of the individuals monitored exceeded 2 rem.

TABLE 3.1
Annual Exposure Information
1997-2010

NRC License Category and Program Code	Year	Number of Licensees	Number of Monitored Individuals	Number of Individuals with Measurable Dose	Collective Dose (person-rem)	Average Measurable Dose (rem)
Industrial Radiography	1997	2	36	12	0.310	0.026
	1998	2	42	33	4.189	0.127
03310	1999	1	9	7	1.585	0.226
03320	2000	12	192	160	65.722	0.411
	2001	9	105	84	27.289	0.325
	2002	14	158	122	35.098	0.288
	2003	25	282	226	109.971	0.487
	2004	24	352	289	159.914	0.553
	2005	26	466	392	247.134	0.630
	2006	44	1,195	1,030	612.518	0.595
	2007	44	1,477	1,313	738.452	0.562
	2008	31	870	748	387.942	0.519
	2009	39	966	723	344.975	0.477
	2010	41	1,045	824	330.055	0.401
Overall Total		**314**	**7,195**	**5,963**	**3,065.154**	**0.514**
Manufacturing and Distribution	1997			*No Data Reported*		
	1998	2	48	12	1.364	0.114
	1999	3	60	16	0.204	0.013
02500	2000	4	87	29	5.320	0.183
03214	2001	7	121	59	6.520	0.111
	2002	125	2,583	1,422	42.922	0.030
	2003	149	3,775	3,044	328.092	0.108
	2004	162	4,073	3,335	321.149	0.096
	2005	158	4,165	3,248	322.476	0.099
	2006	158	4,192	3,352	360.798	0.108
	2007	159	4,179	3,331	349.877	0.105
	2008	156	4,207	3,279	319.200	0.097
	2009	173	4,149	3,299	342.289	0.104
	2010	21	550	333	40.992	0.123
Overall Total		**1,277**	**32,189**	**24,759**	**2,441.203**	**0.099**
Waste Disposal Burial and Service Processing and/or Repackaging	1997	1	5	3	1.508	0.503
	1998	2	33	17	1.954	0.115
	1999	2	42	29	2.869	0.099
	2000	1	5	4	0.874	0.219
03231	2001	1	5	4	0.948	0.237
03234	2002	2	19	13	1.304	0.100
	2003	2	17	10	1.485	0.149
	2004			*No Data Reported*		
	2005	1	5	5	3.120	0.624
	2006	1	5	5	2.601	0.520
	2007	2	13	7	2.263	0.323
	2008	1	5	5	1.717	0.343
	2009	1	3	2	1.712	0.856
	2010	1	3	3	2.681	0.894
Overall Total		**18**	**160**	**107**	**25.036**	**0.234**

TABLE 3.1
Annual Exposure Information (continued)
1997-2010

NRC License Category and Program Code	Year	Number of Licensees	Number of Monitored Individuals	Number of Individuals with Measurable Dose	Collective Dose (person-rem)	Average Measurable Dose (rem)
Irradiators Other Less Than 10000 Curies 03511	1997-1999			*No Data Reported*		
	2000	1	10	5	2.335	0.467
	2001	1	11	6	3.790	0.632
	2002	1	12	6	3.190	0.532
	2003	1	13	4	3.989	0.997
	2004	1	12	4	2.291	0.573
	2005	1	9	4	2.232	0.558
	2006	1	11	5	2.474	0.495
	2007	1	10	4	2.357	0.589
	2008	1	10	4	2.575	0.644
	2009	1	10	4	2.113	0.528
	2010			*No Data Reported*		
Overall Total		**10**	**108**	**46**	**27.346**	**0.594**
Well Logging Byproduct and/or SNM Sealed Sources Only 03111	1997-1999			*No Data Reported*		
	2000	1	68	62	23.724	0.383
	2001	1	83	82	38.305	0.467
	2002	1	69	66	32.331	0.490
	2003	1	60	58	28.011	0.483
	2004	1	60	58	25.012	0.431
	2005	1	65	58	25.165	0.434
	2006	1	68	63	27.714	0.440
	2007	1	68	61	26.623	0.436
	2008	1	70	65	34.384	0.529
	2009	1	68	65	30.503	0.469
	2010	1	68	56	27.050	0.483
Overall Total		**11**	**747**	**694**	**318.822**	**0.459**
Measuring Systems Portable Gauges 03121	1997-1999			*No Data Reported*		
	2000	1	20	18	4.733	0.263
	2001	1	16	15	3.145	0.210
	2002	1	12	11	2.795	0.254
	2003	1	13	12	3.242	0.270
	2004	1	14	11	2.418	0.220
	2005	1	14	13	1.656	0.127
	2006	1	16	16	4.552	0.285
	2007	1	11	9	1.831	0.203
	2008	1	12	12	1.595	0.133
	2009			*No Data Reported*		
	2010	1	15	11	3.900	0.355
Overall Total		**10**	**143**	**128**	**29.867**	**0.233**

TABLE 3.1
Annual Exposure Information (continued)
1997-2010

NRC License Category and Program Code	Year	Number of Licensees	Number of Monitored Individuals	Number of Individuals with Measurable Dose	Collective Dose (person-rem)	Average Measurable Dose (rem)
Veterinary Non-Human 02400	1997-2009			No Data Reported		
	2010	1	4	0	0.000	-
	Overall Total	**1**	**4**	**0**	**0.000**	**-**
Instrument Calibration Service Only - Source > 100 Curies/Other Services 03222 03225	1997-2008			No Data Reported		
	2008	1	3	3	0.000	0.000
	2009	2	36	2	0.414	0.207
	2010	2	37	2	0.247	0.124
	Overall Total	**5**	**76**	**7**	**0.661**	**0.094**
	GRAND TOTALS	**1,646**	**40,622**	**31,704**	**5,908.089**	**0.186**

TABLE 3.2
Summary Distribution of Collective TEDE by License Category
1997-2010

License Category (Number of sites reporting)	Number of Individuals with TEDE in the Ranges (rem)*													Total Number Monitored	Number with Meas. Dose	Total Collective Dose (TEDE) (person-rem)
	No meas.	Meas. <0.1	0.10-0.25	0.25-0.50	0.50-0.75	0.75-1.00	1.00-2.00	2.00-3.00	3.00-4.00	4.00-5.00	5.00-6.00	6.00-12.00	>12			
INDUSTRIAL RADIOGRAPHY																
Fixed Locations (31)	153	58	14	13	4	-	-	-	-	-	-	-	-	242	89	11.712
Temporary Job Sites (283)	1,079	1,897	953	973	571	459	776	197	36	10	1	1	-	6,953	5,874	3,053.442
Total (314)	1,232	1,955	967	936	575	459	776	197	36	10	1	1	-	7,195	5,963	3,065.154
MANUFACTURING AND DISTRIBUTION																
Nuclear Pharmacies (1,254)	7,033	18,451	3,721	1,622	457	172	154	35	14	2	1	-	-	31,662	24,629	2,414.383
Manufacturing & Dist- Other (23)	397	70	22	23	8	3	4	-	-	-	-	-	-	527	130	26.820
Total (1,277)	7,430	18,521	3,743	1,646	466	176	158	36	14	2	1	-	-	32,189	24,759	2,441.203
WASTE DISPOSAL BURIAL AND SERVICE PROCESSING AND/OR REPACKAGING																
Waste Disposal - Burial (2)	26	24	14	-	-	-	-	-	-	-	-	-	-	64	38	2.891
Waste Disposal - Svc Processing(16)	27	22	17	14	7	4	4	1	-	-	-	-	-	96	69	22.145
Total (18)	53	46	31	14	7	4	4	1	-	-	-	-	-	160	107	25.036
IRRADIATORS OTHER LESS THAN 10000 CURIES/IRRADIATORS OTHER GREATER THAN 10000 CURIES																
Irradiators Other < 10000 Curies (10)	62	7	6	12	9	2	9	1	-	-	-	-	-	108	46	27.346
Total (10)	62	7	6	12	9	2	9	1	-	-	-	-	-	108	46	27.346
WELL LOGGING BYPRODUCT AND/OR SNM SEALED SOURCES ONLY/MEASURING SYSTEMS PORTABLE GAUGES																
Well Logging (11)	53	259	119	76	51	71	113	5	-	-	-	-	-	747	694	318.822
Meas Systems-Port gauges (10)	15	43	43	29	6	4	3	-	-	-	-	-	-	143	128	29.067
Total (21)	68	302	162	105	57	75	116	5	-	-	-	-	-	890	822	348.689
VETERINARY NON-HUMAN/INSTRUMENT CALIBRATION SERVICE ONLY - SOURCE > 100 CURIES/OTHER SERVICES																
Veterinary Non-Human (1)	4	-	-	-	-	-	-	-	-	-	-	-	-	4	-	-
Instrument Calibration (2)	69	-	-	-	-	-	-	-	-	-	-	-	-	69	-	-
Other Services (3)	-	5	1	1	-	-	-	-	-	-	-	-	-	7	7	0.661
Total (6)	73	5	1	1	-	-	-	-	-	-	-	-	-	80	7	0.661
GRAND TOTALS	8,918	20,836	4,910	2,763	1,113	716	1,063	239	50	12	2	1	-	40,622	31,704	5,908.089

* Dose values exactly equal to the values separating ranges are reported in the next higher range.

3.3 Summary of Occupational Dose Data by License Category

3.3.1 Industrial Radiography Licenses, Fixed Locations, and Temporary Job Sites

Industrial radiography licenses are issued to allow the use of sealed radioactive materials, usually in exposure devices or cameras, that primarily emit gamma rays for nondestructive testing of pipeline weld joints, steel structures, boilers, aircraft and ship parts, and other high-stress alloy parts. Some firms are licensed to conduct such activities in one location, usually in a permanent facility designed and shielded for radiography; others perform radiography at temporary job sites in the field. The radioisotopes most commonly used are cobalt-60 and iridium-192. Table 3.1 lists the number of licensees that submitted annually from 1997 – 2010; however, only 100 unique radiography licensees reported in that time frame.

Figure 3.1 shows the number of individuals with measurable dose, the total collective dose, and the average measurable dose per individual for both types of industrial radiography licensees from 1997 through 2010. In the last 3 years, the combination of increased individuals with measurable dose plus slightly lower collective TEDE generated a lower average measurable TEDE per individual.

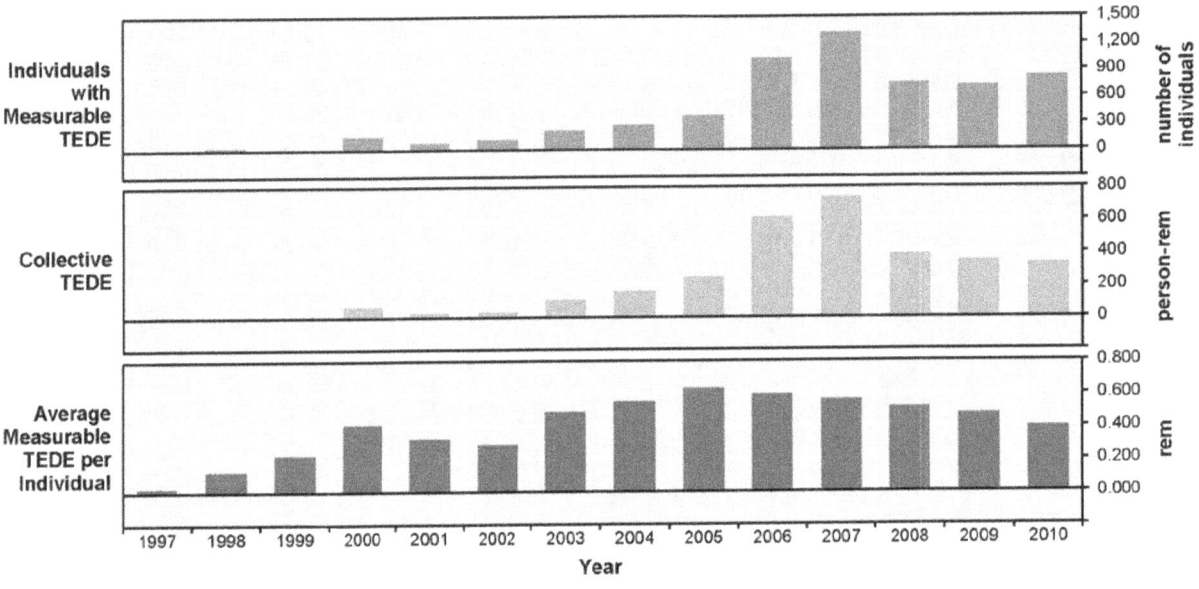

FIGURE 3.1
Annual Values for Industrial Radiography Licensees
1997-2010

Whereas industrial radiographers represent nineteen percent of the reported individuals with a measurable dose (shown in Table 3.2), they received just over half (51.9%) of the total collective dose. In addition, for this time period, 245 workers (4%) received measurable doses greater than or equal to 2 rem. These dose records augment the other 2,120 exposure records for NRC-licensed industrial radiographers that were in the REIRS database for the same time period whose measurable doses also were greater than or equal to 2 rem.

Table 3.3 summarizes the reported data for the two types of industrial radiography licensees for 2008, 2009, and 2010. Table 3.3 also shows that over 99% of the collective dose was attributed to radiographers working at temporary job sites. It is difficult for these individuals to avoid exposure to radiation at temporary job sites in the field. Conditions are not always favorable to provide shielding or to use distance as a means of reducing exposure.

TABLE 3.3
Annual Exposure Information for Industrial Radiographers
2008-2010

Year	Type of License	Number of Licensees	Number of Monitored Individuals	Individuals with Measurable Dose	Collective Dose (person-rem)	Average Measurable Dose (rem)
2008	Fixed Location	4	34	14	1.870	0.13
	Temporary Job Sites	27	836	734	386.072	0.53
	Total	**31**	**870**	**748**	**387.942**	**0.52**
2009	Fixed Location	1	9	1	0.126	0.13
	Temporary Job Sites	38	957	722	344.849	0.48
	Total	**39**	**966**	**723**	**344.975**	**0.48**
2010	Fixed Location	1	10	6	0.083	0.01
	Temporary Job Sites	40	1,035	818	329.972	0.40
	Total	**41**	**1,045**	**824**	**330.055**	**0.40**

3.3.2 Manufacturing and Distribution Licenses

Manufacturing and distribution (M&D) licenses are issued to allow the manufacture and distribution of radionuclides in various forms for a number of diverse purposes. The products are usually distributed to organizations/companies specifically licensed by NRC or an Agreement State. Nuclear pharmacies, the largest licensee contributor, are involved in the compounding and dispensing of radioactive materials for use in nuclear medicine procedures. Table 3.1 lists the number of licensees that submitted annually, with 199 unique manufacturing and distribution licensees reporting from 1998 through 2010.

Figure 3.2 shows the number of individuals with measurable dose, the total collective dose, and the average measurable dose per individual for Type "A" Broad, Type "B" Broad and other, and nuclear pharmacies licensees from 1998 through 2010. In 2010, a large decrease in both the number of individuals reported and the collective TEDE created a slight increase in the average measurable TEDE.

Whereas nuclear pharmacy workers represent nearly 78% of the reported individuals with a measurable dose (shown in Table 3.2), they received 40.9% of the total collective dose. In addition, for this time period, 52 workers (0.2%) received measurable doses greater than or equal to 2 rem. These dose records augment the other 561 exposure records for NRC-licensed nuclear pharmacy workers that were in the REIRS database for the same time period whose measurable doses also were greater than or equal to 2 rem.

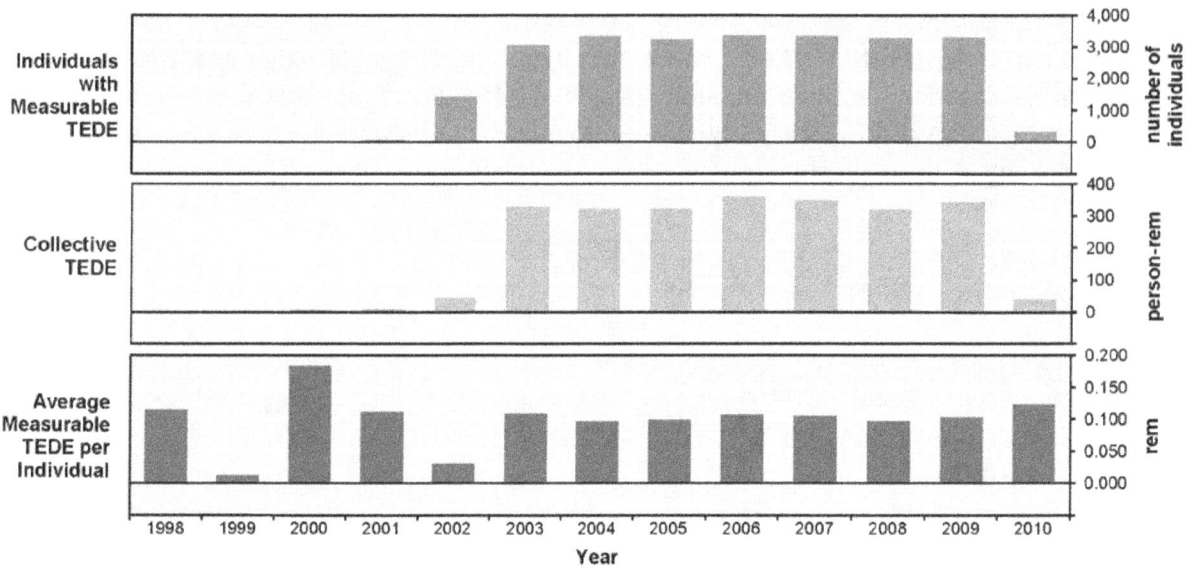

FIGURE 3.2
Annual Values for Manufacturing & Distribution and Nuclear Pharmacies
1998-2010

Table 3.4 summarizes the reported data for manufacturing and distribution licensees for 2008, 2009, and 2010, with the exception of Type "A" Broad licensees that did not submit dose records in any of these 3 years.

TABLE 3.4
Annual Exposure Information for Manufacturing and Distribution
2008-2010

Year	Type of License	Number of Licensees	Number of Monitored Individuals	Individuals with Measurable Dose	Collective Dose (person-rem)	Average Measurable Dose (rem)
2008	M & D - Type "A" Broad			*No Data Reported*		
	M & D - Type "B" Broad and Other	2	43	8	0.770	0.10
	M & D - Nuclear Pharmacies	154	4,164	3,271	318.430	0.10
	Total	**156**	**4,207**	**3,279**	**319.200**	**0.10**
2009	M & D - Type "A" Broad			*No Data Reported*		
	M & D - Type "B" Broad and Other	2	44	12	1.365	0.11
	M & D - Nuclear Pharmacies	171	4,105	3,287	340.924	0.10
	Total	**173**	**4,149**	**3,299**	**342.289**	**0.10**
2010	M & D - Type "A" Broad			*No Data Reported*		
	M & D - Type "B" Broad and Other	2	47	10	4.709	0.47
	M & D - Nuclear Pharmacies	19	503	323	36.283	0.11
	Total	**21**	**550**	**333**	**40.992**	**0.12**

3.3.3 Waste Disposal Burial and Service Processing and/or Repackaging

Waste disposal burial and service processing and/or repackaging licenses are issued to allow the processing or repackaging of radionuclides in various forms for shipment and ultimate disposal. Three unique waste disposal licensees submitted reports from 1997 through 2010.

Figure 3.3 shows the number of individuals with measurable dose, the total collective dose, and the average measurable dose per individual for waste disposal service from 1997 through 2010. As illustrated by Figure 3.3, no data have been received for 2004. Table 3.1 and Figure 3.3 may be updated in the future if data are voluntarily submitted for this specific year.

Table 3.5 summarizes the reported data for waste disposal licensees for 2008, 2009, and 2010.

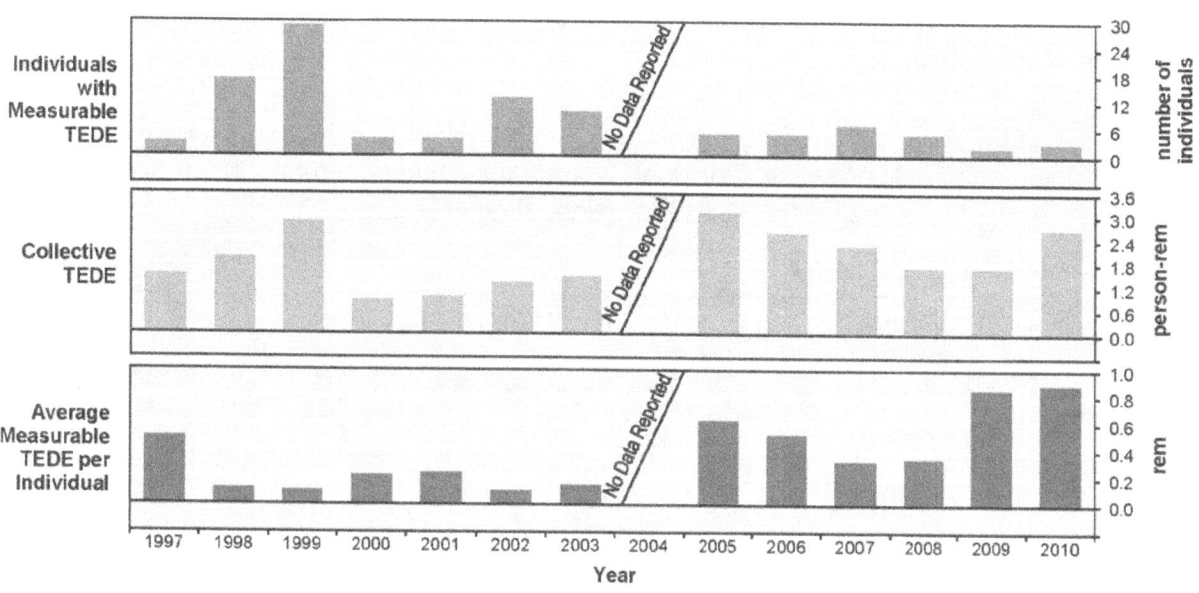

FIGURE 3.3

Annual Values for Waste Disposal Burial and Service Processing and/or Repackaging
1997-2010

TABLE 3.5

Annual Exposure Information for Waste Disposal Burial and Service Processing and/or Repackaging
2008-2010

Year	Type of License	Number of Licensees	Number of Monitored Individuals	Individuals with Measurable Dose	Collective Dose (person-rem)	Average Measurable Dose (rem)
2008	Waste Disposal	1	5	5	1.717	0.34
	Total	1	5	5	1.717	0.34
2009	Waste Disposal	1	3	2	1.712	0.86
	Total	1	3	2	1.712	0.86
2010	Waste Disposal	1	3	3	2.681	0.89
	Total	1	3	3	2.681	0.89

3.3.4 *Irradiators Other Less Than 10000 Curies*

Irradiators other less than 10000 curies licenses are issued for use in irradiation of products, food for human consumption, or research purposes. One single irradiator licensee reported annually from 2000 through 2009.

Figure 3.4 shows the number of individuals with measurable dose, the total collective dose, and the average measurable dose per individual for irradiator licensees from 1997 through 2010. In the past 6 years, the collective TEDE has remained relatively level.

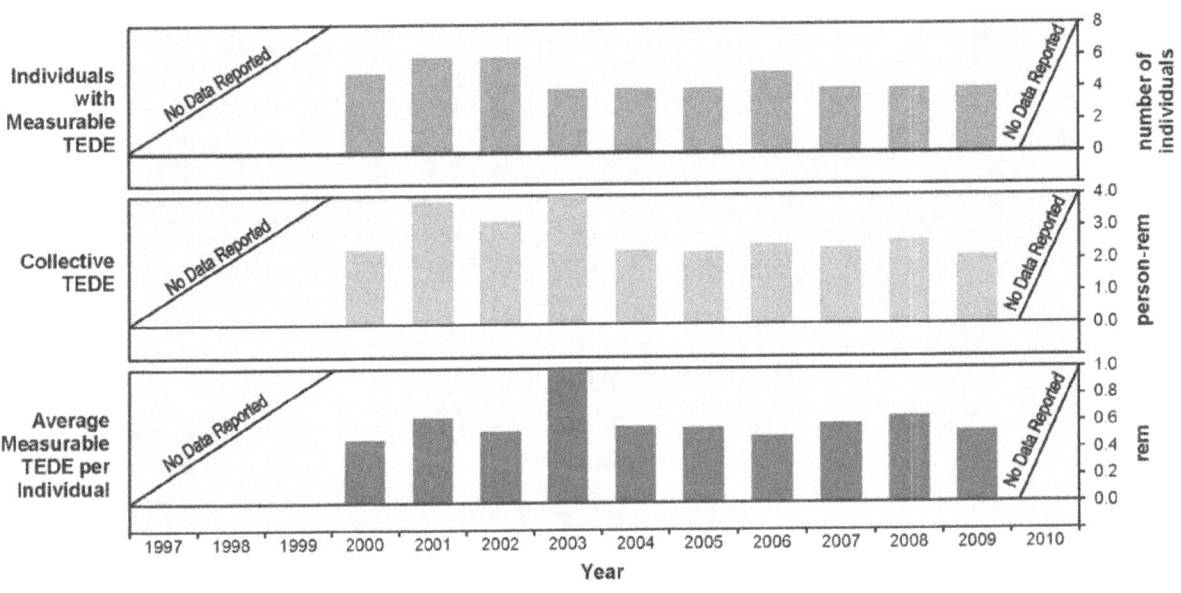

FIGURE 3.4

Annual Values for Irradiators Other Less Than 10000 Curies
1997-2010

Table 3.6 summarizes the reported data for irradiators for 2008, 2009, and 2010.

TABLE 3.6
Annual Exposure Information for Irradiators Other Less than 10000 Curies
2008-2010

Year	Type of License	Number of Licensees	Number of Monitored Individuals	Individuals with Measurable Dose	Collective Dose (person-rem)	Average Measurable Dose (rem)
2008	Irradiators Other < 10000 Curies	1	10	4	2.575	0.64
	Total	**1**	**10**	**4**	**2.575**	**0.64**
2008	Irradiators Other < 10000 Curies	1	10	4	2.113	0.53
	Total	**1**	**10**	**4**	**2.113**	**0.53**
2010	Irradiators Other < 10000 Curies			*No Data Reported*		
	Total	**-**	**-**	**-**	**-**	**-**

3.3.5 Well Logging Byproduct and/or Special Nuclear Material (SNM) Sealed Sources Only and Measuring Systems Portable Gauges

Licenses for sealed byproduct materials are issued for well logging, tracer operations, and research and development. Licenses for portable gauge devices are issued to measure moisture, density, and thickness using sealed sources in a variety of gauge designs. Only two well logging and portable gauge licensees reported from 2000 through 2010.

Figure 3.5 shows the number of individuals with measurable dose, the total collective dose, and the average measurable dose per individual for well logging and portable gauge licensees from 1997 through 2010.

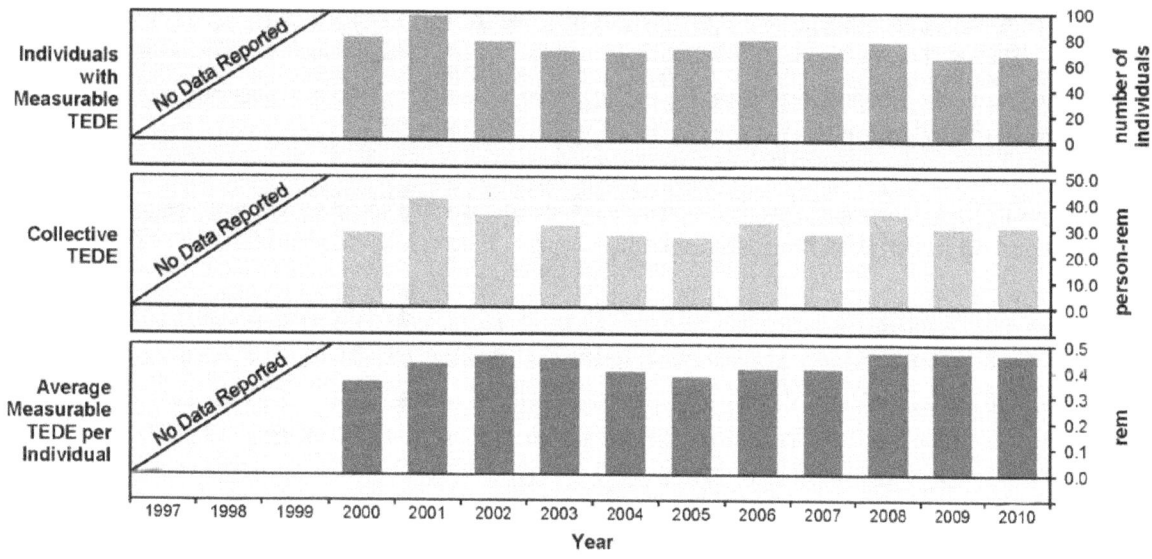

FIGURE 3.5
Annual Values for Well Logging Measuring Systems
1997-2010

Three percent of the reported individuals with measurable dose (shown in Table 3.2) were monitored by well logging, sealed source facilities, and measuring systems and portable gauge facilities, where they received 5.9% of the total collective dose.

Table 3.7 summarizes the reported data for well logging and measuring systems for 2008, 2009, and 2010.

TABLE 3.7
Annual Exposure Information for Well Logging Byproduct and/or SNM Sealed Sources Only
and Measuring Systems Portable Gauges
2008-2010

Year	Type of License	Number of Licensees	Number of Monitored Individuals	Individuals with Measurable Dose	Collective Dose (person-rem)	Average Measurable Dose (rem)
2008	Well Logging and/or SNM Sealed Sources Only	1	70	65	34.384	0.53
	Measuring Systems Portable Gauges	1	12	12	1.595	0.13
	Total	2	82	77	35.979	0.47
2009	Well Logging and/or SNM Sealed Sources Only	1	68	65	30.503	0.47
	Measuring Systems Portable Gauges			No Data Reported		
	Total	1	68	65	30.503	0.47
2010	Well Logging and/or SNM Sealed Sources Only	1	68	56	27.050	0.48
	Measuring Systems Portable Gauges	1	15	11	3.900	0.35
	Total	2	83	67	30.950	0.46

3.3.6 Veterinary Non-Human
Veterinary use includes diagnostic, therapeutic, and research veterinary uses of radioactive drugs and devices. These licenses usually are issued for the treatment of domestic pets and non-food animals. At the present time, no radioactive veterinary drugs have been approved for use in animals intended for the human food supply.

No licensees in this category reported data in 2008 and 2009. Only one licensee in this category reported in 2010 with no measurable dose.

3.3.7 Instrument Calibration Service Only – Source > 100 Curies/Other Services
Instrument calibration services are involved in the calibration of radiation survey and monitoring instruments. Other services possess and use radioactive material for various commercial services, such as teletherapy, industrial gauge servicing, or nuclear laundry. Two licensees reported in these categories from 2008 through 2010.

Figure 3.6 shows the number of individuals with measurable dose, the total collective dose, and the average measurable dose per individual for instrument calibration service only – source > 100 curies, and for other services from 1997 through 2010.

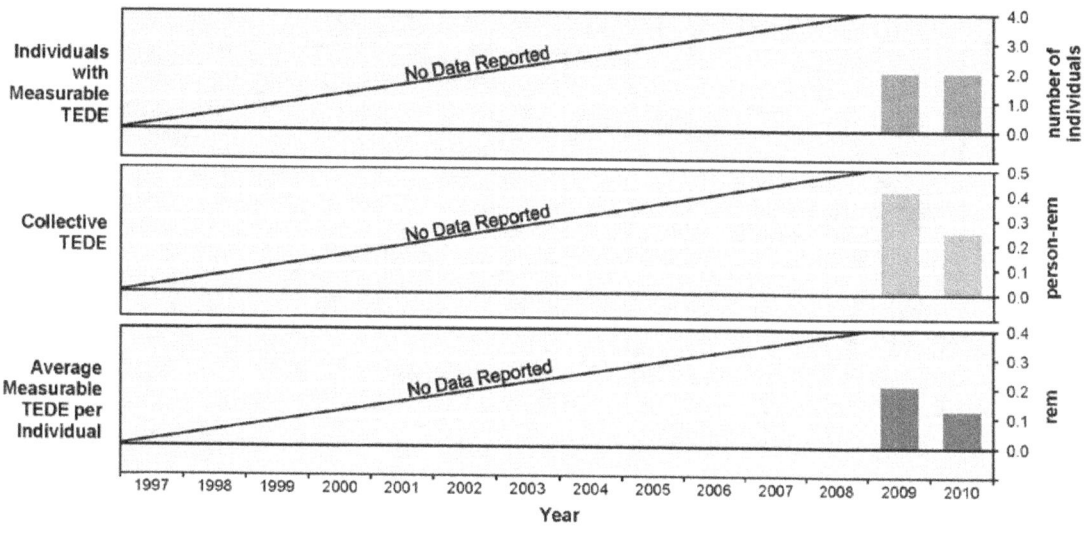

FIGURE 3.6
Annual Values for Instrument Calibration and Other Services
1997-2010

Table 3.8 summarizes the reported data for these licensees for 2008, 2009, and 2010. In 2009 and 2010, licensees for instrument calibration service only – source >100 curies and veterinary non-human submitted all individuals with no measurable exposure. These individuals were monitored and doses were reported as ND or non-detectable, which is represented by zero in Table 3.8. Therefore, no average measurable dose was calculated.

TABLE 3.8
Annual Exposure Information for Veterinary Non-Human, Instrument Calibration Service, Other Services
2008-2010

Year	Type of License	Number of Licensees	Number of Monitored Individuals	Individuals with Measurable Dose	Collective Dose (person-rem)	Average Measurable Dose (rem)
2008	Veterinary Non-Human			No Data Reported		
	Instrument Calibration Service Only - source >100 Curies			No Data Reported		
	Other Services	1	3	0	0.000	-
	Total	**1**	**3**	**0**	**0.000**	**-**
2009	Veterinary Non-Human			No Data Reported		
	Instrument Calibration Service Only - source >100 Curies	1	34	0	0.000	-
	Other Services	1	2	2	0.414	0.21
	Total	**2**	**36**	**2**	**0.414**	**0.21**
2010	Veterinary Non-Human	1	4	0	0.000	-
	Instrument Calibration Service Only - source >100 Curies	1	35	0	0.000	-
	Other Services	1	2	2	0.247	0.12
	Total	**3**	**41**	**2**	**0.247**	**0.12**

4 EXPOSURES TO PERSONNEL IN EXCESS OF REGULATORY LIMITS

4.1 Reporting Categories

Doses in excess of regulatory limits are sometimes referred to as "overexposures." The phrase "doses in excess of regulatory limits" is preferred to overexposures because the latter suggests that an individual has been subjected to an unacceptable biological risk, which may or may not be the case.

The implementation date for the revised 10 CFR Part 20 was January 1, 1994. 10 CFR Part 20 includes requirements for summing internal and external dose equivalents to yield TEDEs and to implement a similar limitation system for organs and tissues (such as the gonads, red bone marrow, bone surfaces, lung, thyroid, and breast). 10 CFR 20.1201 limits the TEDE of individuals to ionizing radiation from licensed material and other sources of radiation within the licensee's control. The annual occupational dose limit for adults is 5 rem.

10 CFR 20.2202 and 10 CFR 20.2203 require that all licensees submit reports of all occurrences involving personnel radiation doses that exceed certain control levels, thus providing for investigations and corrective actions as necessary. Based on the magnitude of the dose, the occurrence may be placed into one of three categories as follows:

1. Category A
   10 CFR 20.2202(a)(1) — a TEDE to any individual of 25 rem or more, a lens dose equivalent of 75 rem or more, or a shallow-dose equivalent to the skin or extremities of 250 rad or more. The Commission must be notified immediately of these events.

2. Category B
   10 CFR 20.2202(b)(1) — In a 24-hour period, the Commission must be notified of the following events: a TEDE to any individual exceeding 5 rem, a lens dose equivalent exceeding 15 rem, or a shallow-dose equivalent to the skin or extremities exceeding 50 rem.

3. Category C
   10 CFR 20.2203 — In addition to the notification required by 10 CFR 20.2202 (Category A or B events), each licensee must submit a written report within 30 days after learning of any of the following occurrences:

   a. Any incident for which notification is required by 10 CFR 20.2202

   b. Doses that exceed the limits in §20.1201, §20.1207, §20.1208, or §20.1301 (for adults, minors, the embryo/fetus of a declared pregnant woman, and the public, respectively) or any applicable limit in the license

   c. Levels of radiation or concentrations of radioactive material that exceed any applicable license limit for restricted areas or that, for unrestricted areas, are in excess of 10 times any applicable limit set forth in 10 CFR Part 20 or in the license (whether or not involving dose of any individual in excess of the limits in §20.1301)

d. For licensees subject to the provisions of the Environmental Protection Agency's generally applicable environmental radiation standards in 40 CFR 190, levels of radiation or releases of radioactive material in excess of those standards or license conditions related to those standards

Exposure events reported as either Category A, B, or C typically undergo an investigation and evaluation process by the licensee, the state department responsible for radiation protection, and NRC inspectors when applicable. Preliminary dose estimates submitted by licensees are often conservatively high and do not represent the final (record) dose assigned for the event. It is, therefore, not uncommon for a dose in excess of a regulatory limit event to be reassessed and the final assigned dose to be categorized as not having been in excess of a regulatory limit. In other cases, the exposure event may not be identified until a later date, such as during the next scheduled audit or inspection of the licensee's event records.

The exposure events summary presented here are for events that occurred between 1997 though 2010. An event that has been reassessed and determined not to be a dose in excess of a regulatory limit is not included in this report. The reader should note that the summary presented here represents a snapshot of the status of events through the end of calendar year 2010.

It is important to note that this summary of events includes only
* Occupational radiation doses in excess of the TEDE regulatory limit
* Events at Agreement State-licensed facilities

It **does not** include
* Medical events as defined in 10 CFR Part 35
* Doses in excess of the regulatory limits to the general public
* Doses in excess of shallow dose equivalent to the maximally exposed extremity (SDE-ME) and shallow dose equivalent to the whole body (SDE-WB)
* Other radiation-related violations, such as high dose-rate areas or effluent limits
* NRC-licensed facilities
* Exposures to dosimeters that, upon evaluation, have been determined to be high dosimeter readings only and are not assigned to an individual as the dose of record by the licensee

4.2 Summary of Occupational Radiation Doses in Excess of NRC Regulatory Limits

From 1997 through 2010, there were three events that resulted in exposures that exceeded the annual TEDE of 5 rem. Two of these incidents were reported to NRC's Nuclear Material Events Database (NMED) and one event was reported to the appropriate Agreement State Radiation

Control Program. Each event underwent an inspection, and the licensees developed a corrective action plan that was reviewed by either the NRC or the appropriate Agreement State Radiation Control Program, to prevent the recurrence of similar events.

In November 2005, a radiation safety officer conducted an update of the radiation exposure reports for their radiography firm. It was noted that for one of the radiographers, a past employer exposure history report could not be located. The past employer history report was eventually found and the radiographer had received 3.918 rem from work conducted at this previous company. When this dose was added to the dose received under his current employer, 1.795 rem, his annual dose was 5.713 rem. The licensee reported this event to the Texas Department of State Health Services and an event description was included in NRC's NMED. This was a category C event with a dose that exceeded the 5 rem TEDE limit.

In June 2007, a positron emission tomography technician received a dose of 5.440 rem while trouble shooting an error on a Chemistry Synthesis Unit (CSU) at a nuclear pharmacy. He made three attempts to inspect and repair the vacuum line behind the CSU without checking his pocket dosimeter. When this dose was added to his current dose, it brought his annual dose to 5.957 rem. The licensee reported this event to the Washington Department of Health. This was a category C event with a dose that exceeded the 5 rem TEDE limit.

In 2008, a radiography licensee received an exposure report for one of its terminated employees. The exposure report indicated that the employee received 8.212 rem. The radiography licensee conducted a review of the employee's exposure reports and tried to contact the employee to obtain additional information. Since the radiography licensee was unsuccessful in contacting the former employee, the radiography licensee reported this overexposure to the California Health and Human Services Agency and an event description was included in NRC's NMED. Since this employee conducted work at other facilities, it was determined that the total annual dose was 8.367 rem. This was a Category C event with a dose that exceeded the 5 rem TEDE limit.

4.3 Summary of Annual Dose Distributions for Certain Agreement State Licensees

The dose distribution shown in Table 3.2 reflects the number of individuals in each dose range as reported by each licensee. Each licensee reports the radiation exposure records for the individuals that worked at their facility. Individuals may work at more than one licensee during the year, and this is not reflected in the dose distribution compiled per licensee. Individuals monitored at more than one licensee are referred to as "transient" individuals. Transient individuals in the nuclear power industry are common and have a significant impact on the dose distribution. To determine the impact that transients have on the dose distribution for the Agreement State licensees, the doses are summed for each individual during the year (independent of location) to determine the annual value for each individual. Of particular interest are the individuals that may have exceeded a dose limit by receiving doses at multiple facilities.

Five rem is the regulatory dose limit for TEDE in 10 CFR Part 20.1201. Agreement States enforce this same regulatory limit. Two rem is commonly used as an administrative control level to ensure that doses do not approach the 5 rem limit and also to keep doses ALARA. The NRC is assessing the possibility of reducing the dose limit from 5 rem per year to 2 rem per year. Since a change to NRC's 5 rem per year regulatory limit could impact Agreement State licensees, assessing the number of individuals that exceed 2 rem per year at Agreement State licensees is included in this report.

Table 4.1 gives a summary of the individuals reported by Agreement State licensees and shows the number of reported individuals exceeding 2 rem and 5 rem per year from 1997 to 2010. Table 4.1 also shows that for 11 of the past 14 years, the percentage of individuals with less than 2 rem has been greater than 99%. The number of individuals receiving an annual dose greater than or equal to 5 rem has been less than 0.02% since 2005 and has been decreasing over the past 3 years. No individual monitored at any of the Agreement State licensee categories included in this report received a dose above the 5 rem annual TEDE limit during the past 2 years. It should be noted that this report contains only those Agreement State licensees that voluntarily reported their data to REIRS. Therefore, this data is only a representative sample of the dose information from all Agreement State licensees.

It is not possible to assess the impact on Agreement State licensees if the regulatory limit is changed to 2 rem because according to Table 4.1, few individuals have exceeded 2 rem. In addition, the data voluntarily reported by Agreement State licensees represents ~2% of all Agreement State licensees. As noted above, this data set is a representative sample and a generic assessment may not be applicable across the broad range of Agreement State licensee activities. However, as a general assessment, if the current regulatory limit were reduced from 5 rem to 2 rem, licensees may need to improve radiation protection practices, revise procedures, and possibly utilize more workers in certain tasks in order to distribute the dose among the worker population. A combination of these approaches could lessen the impact of a potential change in the regulatory limit, but there may still be challenges in reducing dose further in some occupational situations due to limitations in the availability of skilled workers that could perform certain tasks.

As shown in the Corrected Number column of Table 4.1, the impact of transient individuals to the dose received is minimal. Over the past 14 years, just over 1% of the total number of reported individuals may have been counted more than once because they worked at more than one facility during the calendar year.

TABLE 4.1

Summary of Annual Distributions for Certain Agreement State Licensees
1997-2010

Year	Total Number of Monitored Individuals		Individuals with Doses **				Individuals with Doses >12 rem **
			< 2 rem	> 2 rem	< 5 rem	> 5 rem	
	Reported Number	Corrected Number *	%	Number	%	Number	
1997	41	41	100.0%	0	100.0%	0	0
1998	123	101	100.0%	0	100.0%	0	0
1999	111	111	100.0%	0	100.0%	0	0
2000	382	381	99.2%	3	100.0%	0	0
2001	357	355	99.4%	2	100.0%	0	0
2002	2,850	2,847	99.9%	2	100.0%	0	0
2003	4,160	4,114	99.6%	17	100.0%	0	0
2004	4,511	4,463	99.6%	16	100.0%	0	0
2005	4,724	4,665	99.1%	42	99.98%	1	0
2006	5,487	5,429	98.8%	67	100.0%	0	0
2007	5,752	5,688	98.8%	68	99.98%	1	0
2008	5,179	5,125	99.3%	35	99.98%	1	0
2009	5,233	5,177	99.4%	31	100.0%	0	0
2010	1,723	1,718	98.7%	22	100.0%	0	0

*   *This column lists the actual number of individuals who may have been counted more than once because they worked at more than one facility during the calendar year.*

** *Data for 1997–2010 are based on the distribution of individual doses after adjusting for the multiple counting of transient individuals.*

4.4 Maximum Occupational Radiation Doses Below NRC Regulatory Limits

Certain researchers have expressed an interest in a listing of the maximum doses received at licensed facilities that do not exceed NRC regulatory limits. This information allows an examination of these doses and could possibly provide insights on where certain improvements can be made regarding licensees' radiation protection programs.

Table 4.2 shows the maximum doses for each dose category required to be reported to the NRC. In addition, the number of doses in certain dose ranges is shown to reflect the number of doses that approach NRC regulatory limits. As shown in Table 4.2, few doses exceed half of the NRC occupational annual limits. The Agreement State occupational data analyzed over the 14-year period for this report indicate that 16 individuals exceeded 95% of the extremity dose limit and 3 individuals exceeded 95% of the 5 rem TEDE limit.

TABLE 4.2

Maximum Occupational Exposures* for Each Exposure Category
1997-2010

Dose Category**	Annual Does Limit 10CFR20***	Maximum Dose Reported (rem)	Max Dose Percent of the Limit	Number of Individuals with Measurable Dose	Number of Individuals >25% of the Limit	Number of Individuals >50% of the Limit	Number of Individuals >75% of the Limit	Number of Individuals >95% of the Limit	Number of Individuals > Limit
SDE-ME	50 rem	60.560	121%	17,874	2763	404	57	16	10
SDE-WB	50 rem	13.412	27%	31,177	2	0	0	0	0
LDE	15 rem	8.856	59%	30,302	30	2	0	0	0
CEDE		2.086		2,449					
CDE		17.197		3,136					
DDE		8.367		31,121					
TEDE	5 rem	8.367	167%	31,379	942	147	24	3	3
TODE	50 rem	17.362	35%	30,516	3	0	0	0	0

* *Numbers have been adjusted for the multiple reporting of transient individuals.*

** *SDE-ME = shallow dose equivalent to the maximally exposed extremity*

   SDE-WB = shallow dose equivalent to the whole body

   LDE = lens dose equivalent

   CEDE = committed effective dose equivalent

   CDE = committed dose equivalent

   DDE = deep dose equivalent

   TEDE = total effective dose equivalent

   TODE = total organ dose equivalent

*** *Shaded boxes represent dose categories that do not have specific dose limits defined in 10 CFR Part 20.*

5 CONCLUSIONS

The occupational radiation exposure data submitted in response to the NRC Letter proved to be a valuable addition to the REIRS database. This data provided insight into the non-NRC licensees to better inform the NRC on possible impacts of changing the regulations to coincide with ICRP 103. In particular, the dose records voluntarily submitted in response to the NRC letter added 37,491 records to the REIRS database in nine licensee categories.

From 1997 to 2010, the Agreement State dose records greater than or equal to 2 rem for industrial radiography licensees totaled 4% (245) of the number with measurable dose, while the dose records greater than or equal to 2 rem for manufacturing and distribution licensees totaled 0.2% (52) of the number with measurable dose. While the manufacturing and distribution licensees have more individuals with measurable dose than industrial radiography licensees, the percentage of individuals exceeding 2 rem for industrial radiography is larger.

The Agreement State data included in this report show that the percentage of individuals with measurable dose that exceed 2 rem at Agreement State licensees was 1%.   The percentage for all NRC licensees was 0.6%. With nearly 20,000 licenses issued in Agreement States, this report analyzed dose records from only 312 (2%) Agreement State licensees. As a result of this report, it is expected that there will be an increase in the number of Agreement State licensees who voluntarily submit their data to REIRS for future analysis.

6 REFERENCES

1. SECY-08-0197. *"Options to Revise Radiation Protection Regulations and Guidance with Respect to the 2007 Recommendations of the International Commission on Radiological Protection,"* dated December 18, 2008 (ML083360582).

2. International Commission on Radiological Protection Publication 103. *The 2007 Recommendations of the International Commission on Radiological Protection*, Annals of the ICRP Volume 37 Nos. 2-4, 2007.

3. SRM-SECY-08-0197. *"Options to Revise Radiation Protection Regulations and Guidance with Respect to the 2007 Recommendations of the International Commission on Radiological Protection,"* dated April 2, 2009 (ML090920103).

4. NRC Letter. *Request to Provide Occupational Radiation Dose Data from Industrial Radiography and Nuclear Pharmacy Licensees* (ML10210039), dated August, 2010.

5. Agreement State Program. http://www.nrc.gov/about-nrc/state-tribal/agreement-states.html.

6. The Atomic Energy Act of 1954, as amended, dated August 30, 1954.

7. U. S. Nuclear Regulatory Commission. *Information Digest 2009-2010, USNRC Report NUREG-1350, Volume 21, August 2009.*

8. Organization of Agreement States. http://www.agreementstates.org/.

9. Conference of Radiation Control Programs, Inc. http://www.crcpd.org/about/about.aspx.

10. U. S. Nuclear Regulatory Commission. *Management Directive 5.9, "Adequacy and Compatibility of Agreement State Programs,"* February 1998

Appendix A

LIST OF AGREEMENT STATE LICENSEES, 1997 - 2010

APPENDIX A

Table A - Annual TEDE for Agreement State Licensees

1997

PROGRAM CODE - LICENSEE NAME	LICENSE#	No. Meas. Exposure	Number of Individuals with Whole Body Doses in the Ranges (rems)*												Total Number Monitored	Number with Meas. Dose	Total Collective TEDE (person-rem)	Average Meas. TEDE (rem)
			Meas. <0.10	0.10- 0.25	0.25- 0.50	0.50- 0.75	0.75- 1.00	1.00- 2.00	2.00- 3.00	3.00- 4.00	4.00- 5.00	5.00- 6.00	6.00- 12.00	>12.0				
WASTE DISPOSAL SERVICE PROCESSING AND/OR REPACKAGING - 03234																		
ENVIRONMENTAL MANAGEMENT & CONTROLS	3546-50	2	-	-	2	-	1	-	-	-	-	-	-	-	5	3	1.508	0.503
Total	1	2	-	-	2	-	1	-	-	-	-	-	-	-	5	3	1.508	0.503
INDUSTRIAL RADIOGRAPHY – TEMPORARY JOB SITE – 03320																		
MQS INSPECTION, INC.	IL-01136-01	1	-	-	-	-	-	-	-	-	-	-	-	-	1	-	-	-
EG&G FLORIDA, INC.	FL-1219-1	23	12	-	-	-	-	-	-	-	-	-	-	-	35	12	0.310	0.026
Total	2	24	12	-	-	-	-	-	-	-	-	-	-	-	36	12	0.310	0.026

APPENDIX A

Table A - Annual TEDE for Agreement State Licensees
1998

PROGRAM CODE - LICENSEE NAME	LICENSE#	No Meas. Exposure	Number of Individuals with Whole Body Doses in the Ranges (rems)*												Total Number Monitored	Number with Meas. Dose	Total Collective TEDE (person-rem)	Average Meas. TEDE (rem)
			Meas. <0.10	0.10-0.25	0.25-0.50	0.50-0.75	0.75-1.00	1.00-2.00	2.00-3.00	3.00-4.00	4.00-5.00	5.00-6.00	6.00-12.00	>12.0				
MANUFACTURING AND DISTRIBUTION – NUCLEAR PHARMACIES – 02500																		
EASTERN ISOTOPES	MD-03-068-01	14	5	1	2	1	-	-	-	-	-	-	-	-	23	9	1.350	0.150
WYLE LABORATORIES	FL2953-1	22	3	-	-	-	-	-	-	-	-	-	-	-	25	3	0.014	0.005
Total	2	36	8	1	2	1	-	-	-	-	-	-	-	-	48	12	1.364	0.114
WASTE DISPOSAL (BURIAL) – 03231																		
US ECOLOGY	WN-I019-2	14	5	8	-	-	-	-	-	-	-	-	-	-	27	13	1.355	0.104
Total	1	14	5	8	-	-	-	-	-	-	-	-	-	-	27	13	1.355	0.104
WASTE DISPOSAL SERVICE PROCESSING AND/OR REPACKAGING - 03234																		
ENVIRONMENTAL MANAGEMENT & CONTROLS	3546-50	2	2	1	1	-	-	-	-	-	-	-	-	-	6	4	0.599	0.150
Total	1	2	2	1	1	-	-	-	-	-	-	-	-	-	6	4	0.599	0.150
INDUSTRIAL RADIOGRAPHY – TEMPORARY JOB SITE – 03320																		
EG&G FLORIDA, INC	FL-1219-1	8	25	-	-	-	-	-	-	-	-	-	-	-	33	25	0.069	0.003
WESTEX COMPANY	CA 5324-56	1	2	1	2	1	-	2	2	-	-	-	-	-	9	8	4.120	0.515
Total	2	9	27	1	2	1	-	2	2	-	-	-	-	-	42	33	4.189	0.127

APPENDIX A

Table A - Annual TEDE for Agreement State Licensees
1999

PROGRAM CODE - LICENSEE NAME	LICENSE#	No Meas. Exposure	Meas. <0.10	0.10-0.25	0.25-0.50	0.50-0.75	0.75-1.00	1.00-2.00	2.00-3.00	3.00-4.00	4.00-5.00	5.00-6.00	6.00-12.00	>12.0	Total Number Monitored	Number with Meas. Dose	Total Collective TEDE (person-rem)	Average Meas. TEDE (rem)
MANUFACTURING AND DISTRIBUTION – NUCLEAR PHARMACIES – 02500																		
EASTERN ISOTOPES	IL-02074-01	5	-	-	-	-	-	-	-	-	-	-	-	-	5			
WYLE LABORATORIES	FL-2953-1	11	15	-	-	-	-	-	-	-	-	-	-	-	26	15	0.129	0.009
Total	2	16	15	-	-	-	-	-	-	-	-	-	-	-	31	15	0.129	0.009
MANUFACTURING AND DISTRIBUTION - OTHER – 03214																		
HOCHIKI AMERICA CORP	2090-30	28	1	-	-	-	-	-	-	-	-	-	-	-	29	1	0.075	0.075
Total	1	28	1	-	-	-	-	-	-	-	-	-	-	-	29	1	0.075	0.075
WASTE DISPOSAL (BURIAL) - 03231																		
US ECOLOGY	WN-I019-2	12	19	6	-	-	-	-	-	-	-	-	-	-	37	25	1.536	0.061
Total	1	12	19	6	-	-	-	-	-	-	-	-	-	-	37	25	1.536	0.061
WASTE DISPOSAL SERVICE PROCESSING AND/OR REPACKAGING - 03234																		
ENVIRONMENTAL MANAGEMENT & CONTROLS	3546-50	1	2	-	1	-	1	-	-	-	-	-	-	-	5	4	1.333	0.333
Total	1	1	2	-	1	-	1	-	-	-	-	-	-	-	5	4	1.333	0.333
INDUSTRIAL RADIOGRAPHY – TEMPORARY JOB SITE – 03320																		
WESTEX COMPANY	CA 5324-56	2	2	3	1	1	-	-	-	-	-	-	-	-	9	7	1.585	0.226
Total	1	2	2	3	1	1	-	-	-	-	-	-	-	-	9	7	1.585	0.226

Number of Individuals with Whole Body Doses in the Ranges (rems)*

APPENDIX A

Table A – Annual TEDE for Agreement State Licensees
2000

PROGRAM CODE - LICENSEE NAME	LICENSE#	No Meas. Exposure	Meas. <0.10	0.10-0.25	0.25-0.50	0.50-0.75	0.75-1.00	1.00-2.00	2.00-3.00	3.00-4.00	4.00-5.00	5.00-6.00	6.00-12.00	>12.0	Total Number Monitored	Number with Meas. Dose	Total Collective TEDE (person-rem)	Average Meas. TEDE (rem)
MANUFACTURING AND DISTRIBUTION – NUCLEAR PHARMACIES – 02500																		
MORAVEK BIOCHEMICALS, INC.	2960-30	6	4	2	2	1	-	-	-	-	-	-	-	-	15	9	1.799	0.200
WYLE LABORATORIES	FL-2953-1	22	1	-	-	-	-	-	-	-	-	-	-	-	23	1	0.011	0.011
Total	2	28	5	2	2	1	-	-	-	-	-	-	-	-	38	10	1.810	0.181
WELL LOGGING BYPRODUCT AND/OR SNM SEALED SOURCES ONLY - 03111																		
ISOTOPE PRODUCTS LABS	1509-19	6	21	14	9	5	8	5	-	-	-	-	-	-	68	62	23.724	0.383
Total	1	6	21	14	9	5	8	5	-	-	-	-	-	-	68	62	23.724	0.383
MEASURING SYSTEMS PORTABLE GAUGES - 03121																		
CPN INTERNATIONAL, INC	1100-07	2	7	4	3	2	2	-	-	-	-	-	-	-	20	18	4.733	0.263
Total	1	2	7	4	3	2	2	-	-	-	-	-	-	-	20	18	4.733	0.263
MANUFACTURING AND DISTRIBUTION – OTHER - 03214																		
HOCHIKI AMERICA CORP	2090-30	23	-	-	-	-	-	-	-	-	-	-	-	-	23	-	-	-
J. L. SHEPHERD AND ASSOCIATES	CA 1777-19	7	8	8	2	-	1	-	-	-	-	-	-	-	26	19	3.510	0.185
Total	2	30	8	8	2	-	1	-	-	-	-	-	-	-	49	19	3.510	0.185
WASTE DISPOSAL SERVICE PROCESSING AND/OR REPACKAGING - 03234																		
ENVIRONMENTAL MANAGEMENT & CONTROLS	3546-50	1	1	1	2	-	-	-	-	-	-	-	-	-	5	4	0.874	0.219
Total	1	1	1	1	2	-	-	-	-	-	-	-	-	-	5	4	0.874	0.219
INDUSTRIAL RADIOGRAPHY – FIXED LOCATION – 03310																		
SONGS RADIOGRAPHY GROUP	CA-5244-30	8	-	-	-	-	-	-	-	-	-	-	-	-	8	-	-	-
THE FLOWSERVE CORPORATION	OH-033105800000	-	4	-	-	-	-	-	-	-	-	-	-	-	4	4	0.094	0.024
Total	2	8	4	-	-	-	-	-	-	-	-	-	-	-	12	4	0.094	0.024
INDUSTRIAL RADIOGRAPHY – TEMPORARY JOB SITE – 03320																		
ALPHA-OMEGA SERVICES, INC.	3925-19	2	2	5	4	1	2	2	-	-	-	-	-	-	18	16	7.331	0.458
ANVIL CORPORATION	VVN-IR031-1	6	13	17	16	9	9	7	1	-	-	-	-	-	78	72	34.342	0.477
CONSTRUCTION MATERIALS TESTING, INC.	0799-07	3	1	-	1	2	1	-	-	-	-	-	-	-	8	5	3.865	0.773
INTERNATIONAL INSPECTION, INC.	5046-19	4	3	2	2	1	-	-	-	-	-	-	-	-	12	8	1.675	0.209
NDE SERVICES, INC.	406-01	3	5	1	-	2	-	-	-	-	-	-	-	-	12	9	2.385	0.265
NDT LABORATORIES, INC.	1651-43	1	4	-	1	-	-	-	-	-	-	-	-	-	6	5	0.473	0.095
TC INSPECTION INC	5299-07	5	6	3	5	2	1	1	-	-	-	-	-	-	23	18	6.755	0.375
TESTING ENGINEERS, INC.	1898-01	-	1	2	1	-	1	-	-	-	-	-	-	-	5	5	1.610	0.322
WESTERN INDUSTRIAL X-RAY	4424-48	-	4	2	3	-	2	1	-	-	-	-	-	-	12	12	4.912	0.409
WESTERN X-RAY CORPORATION	5324-56	-	2	1	1	1	-	-	-	-	-	-	-	-	6	6	2.280	0.380
Total	10	24	41	33	34	18	16	11	3	-	-	-	-	-	180	156	65.628	0.421
IRRADIATORS - OTHER LESS THAN 10000 CURIES - 03511																		
INDUSTRIAL NUCLEAR CO., INC	2229-01	5	1	1	1	-	2	-	-	-	-	-	-	-	10	5	2.335	0.467
Total	1	5	1	1	1	-	2	-	-	-	-	-	-	-	10	5	2.335	0.467

Number of Individuals with Whole Body Doses in the Ranges (rems)

APPENDIX A

Table A - Annual TEDE for Agreement State Licensees
2001

PROGRAM CODE - LICENSEE NAME	LICENSE#	No Meas. Exposure	Meas. <0.10	0.10-0.26	0.25-0.50	0.50-0.76	0.75-1.00	1.00-2.00	2.00-3.00	3.00-4.00	4.00-5.00	5.00-6.00	6.00-12.00	>12.0	Total Number Monitored	Number with Meas. Dose	Total Collective TEDE (person-rem)	Average Meas. TEDE (rem)
MANUFACTURING AND DISTRIBUTION – NUCLEAR PHARMACIES – 02500																		
GAMMA PLUS - DENVER	987-01	2	-	3	1	-	-	-	-	-	-	-	-	-	6	4	0.726	0.182
GE HEALTHCARE - SAN JOSE	4811-43	1	9	1	-	-	-	-	-	-	-	-	-	-	11	10	0.433	0.043
MALLINCKRODT. INC	859-01	-	11	3	-	-	-	-	-	-	-	-	-	-	14	14	0.664	0.047
MORAVEK BIOCHEMICALS, INC.	2960-30	6	5	-	4	1	-	-	-	-	-	-	-	-	16	10	2.123	0.212
WYLE LABORATORIES	FL-2953-1	14	6	-	-	-	-	-	-	-	-	-	-	-	20	6	0.161	0.027
Total	5	23	31	7	5	1	-	-	-	-	-	-	-	-	67	44	4.107	0.093
WELL LOGGING BYPRODUCT AND/OR SNM SEALED SOURCES ONLY - 03111																		
ISOTOPE PRODUCTS LABS	1509-19	1	29	14	11	5	7	15	1	-	-	-	-	-	83	82	38.305	0.467
Total	1	1	29	14	11	5	7	15	1	-	-	-	-	-	83	82	38.305	0.467
MEASURING SYSTEMS PORTABLE GAUGES - 03121																		
CPN INTERNATIONAL, INC.	1100-07	1	4	6	4	1	-	-	-	-	-	-	-	-	16	15	3.145	0.210
Total	1	1	4	6	4	1	-	-	-	-	-	-	-	-	16	15	3.145	0.210
MANUFACTURING AND DISTRIBUTION – OTHER – 03214																		
HOCHIKI AMERICA CORP	2090-30	25	1	-	-	1	-	-	-	-	-	-	-	-	27	2	0.543	0.272
J. L. SHEPHERD AND ASSOCIATES	CA 1777-19	14	10	1	1	-	-	1	-	-	-	-	-	-	27	13	1.870	0.144
Total	2	39	11	1	1	1	-	1	-	-	-	-	-	-	54	15	2.413	0.161
WASTE DISPOSAL SERVICE PROCESSING AND/OR REPACKAGING - 03234																		
ENVIRONMENTAL MANAGEMENT & CONTROLS	3546-50	1	-	1	3	-	-	-	-	-	-	-	-	-	5	4	0.948	0.237
Total	1	1	-	1	3	-	-	-	-	-	-	-	-	-	5	4	0.948	0.237
INDUSTRIAL RADIOGRAPHY – FIXED LOCATION - 03310																		
SONGS - RADIOGRAPHY GROUP	CA-5244-30	8	-	-	-	-	-	-	-	-	-	-	-	-	8	-	-	-
Total	1	8	-	-	-	-	-	-	-	-	-	-	-	-	8	-	-	-

APPENDIX A
Table A - Annual TEDE for Agreement State Licensees
2001 (continued)

PROGRAM CODE - LICENSEE NAME	LICENSE#	No Meas. Exposure	Number of Individuals with Whole Body Doses in the Ranges (rems)*												Total Number Monitored	Number with Meas. Dose	Total Collective TEDE (person-rem)	Average Meas. TEDE (rem)
			Meas. <0.10	0.10- 0.25	0.25- 0.50	0.50- 0.75	0.75- 1.00	1.00- 2.00	2.00- 3.00	3.00- 4.00	4.00- 5.00	5.00- 6.00	6.00- 12.00	>12.0				
INDUSTRIAL RADIOGRAPHY – TEMPORARY JOB SITE – 03320																		
ALPHA-OMEGA SERVICES, INC.	3925-19	3	16	4	3	-	-	-	-	-	-	-	-	-	26	23	2.024	0.088
CONSTRUCTION MATERIALS TESTING, INC.	0799-C7	-	1	-	1	1	2	1	-	-	-	-	-	-	6	6	4.330	0.722
INTERNATIONAL INSPECTION	5046-19	2	2	1	5	-	-	-	-	-	-	-	-	-	10	8	2.140	0.268
NDT LABORATORIES, INC.	1651-43	5	1	-	1	1	-	-	-	-	-	-	-	-	8	3	1.080	0.360
TC INSPECTION, INC.	5299-07	2	7	5	7	2	2	1	-	-	-	-	-	-	26	24	7.800	0.325
TESTING ENGINEERS, INC	1898-01	-	-	2	-	2	-	-	-	-	-	-	-	-	4	4	1.535	0.384
WESTERN INDUSTRIAL X-RAY	4424-43	1	1	2	3	-	1	2	-	-	-	-	-	-	10	9	5.535	0.615
WESTERN X-RAY CORPORATION	5324-56	-	4	1	1	-	-	-	1	-	-	-	-	-	7	7	2.845	0.406
Total	8	13	32	15	21	6	5	4	1	-	-	-	-	-	97	84	27.289	0.325
IRRADIATORS - OTHER - LESS THAN 10000 CURIES - 03511																		
INDUSTRIAL NUCLEAR CO., INC.	2229-01	5	-	1	2	2	-	1	-	-	-	-	-	-	11	6	3.790	0.632
Total	1	5	-	1	2	2	-	1	-	-	-	-	-	-	11	6	3.790	0.632
No Program Code																		
BWX TECHNOLOGIES, INC.	03310760006	3	13	-	-	-	-	-	-	-	-	-	-	-	16	13	0.112	0.009
Total	1	3	13	-	-	-	-	-	-	-	-	-	-	-	16	13	0.112	0.009

37

APPENDIX A
Table A - Annual TEDE for Agreement State Licensees
2002

MANUFACTURING AND DISTRIBUTION – NUCLEAR PHARMACIES – 02500

PROGRAM CODE - LICENSEE NAME	LICENSE#	No Meas. Exposure	Meas. <0.10	0.10-0.25	0.25-0.50	0.50-0.75	0.75-1.00	1.00-2.00	2.00-3.00	3.00-4.00	4.00-5.00	5.00-6.00	6.00-12.00	>12.0	Total Number Monitored	Number with Meas. Dose	Total Collective TEDE (person-rem)	Average Meas. TEDE (rem)
CARDINAL HEALTH - AL	1068	5	6	-	-	-	-	-	-	-	-	-	-	-	11	6	0.069	0.012
CARDINAL HEALTH - AL	1168	1	13	1	-	-	-	-	-	-	-	-	-	-	15	14	0.506	0.036
CARDINAL HEALTH - AR	ARK-642-02500	14	21	-	-	-	-	-	-	-	-	-	-	-	35	21	0.262	0.012
CARDINAL HEALTH - AZ	01-084	-	8	-	-	-	-	-	-	-	-	-	-	-	8	8	0.248	0.031
CARDINAL HEALTH - AZ	07-123	19	24	-	-	-	-	-	-	-	-	-	-	-	43	24	0.369	0.015
CARDINAL HEALTH - CA	2891-37	1	15	-	-	-	-	-	-	-	-	-	-	-	16	15	0.519	0.035
CARDINAL HEALTH - CA	3317-19	9	4	-	-	-	-	-	-	-	-	-	-	-	13	4	0.016	0.004
CARDINAL HEALTH - CA	3426-43	3	5	-	-	-	-	-	-	-	-	-	-	-	8	5	0.037	0.007
CARDINAL HEALTH - CA	3469-1	-	4	-	-	-	-	-	-	-	-	-	-	-	4	4	0.095	0.024
CARDINAL HEALTH - CA	3469-2	2	2	2	-	-	-	-	-	-	-	-	-	-	6	4	0.339	0.085
CARDINAL HEALTH - CA	3673-34	24	13	-	-	-	-	-	-	-	-	-	-	-	37	13	0.132	0.010
CARDINAL HEALTH - CA	3822-19	19	25	-	-	-	-	-	-	-	-	-	-	-	44	25	0.313	0.013
CARDINAL HEALTH - CA	3832-01	10	18	-	-	-	-	-	-	-	-	-	-	-	28	18	0.385	0.021
CARDINAL HEALTH - CA	4905-10	7	12	-	-	-	-	-	-	-	-	-	-	-	19	12	0.084	0.007
CARDINAL HEALTH - CA	4999-30	1	30	-	-	-	-	-	-	-	-	-	-	-	31	30	0.239	0.008
CARDINAL HEALTH - CA	5218-36	1	14	1	-	-	-	-	-	-	-	-	-	-	16	15	0.546	0.036
CARDINAL HEALTH - CA	5905-15	-	5	-	-	-	-	-	-	-	-	-	-	-	5	5	0.023	0.005
CARDINAL HEALTH - CA	5910-50	2	3	-	-	-	-	-	-	-	-	-	-	-	5	3	0.060	0.020
CARDINAL HEALTH - CA	6321-45	1	1	-	-	-	-	-	-	-	-	-	-	-	2	1	0.001	0.001
CARDINAL HEALTH - CA	6891-33	3	5	-	-	-	-	-	-	-	-	-	-	-	8	5	0.185	0.037
CARDINAL HEALTH - CA	6924-36	-	1	2	1	-	-	-	-	-	-	-	-	-	4	4	0.833	0.208
CARDINAL HEALTH - CA	6925-19	1	5	-	-	-	-	-	-	-	-	-	-	-	6	5	0.116	0.023
CARDINAL HEALTH - CO	392-01	2	1	-	-	-	-	-	-	-	-	-	-	-	3	1	0.001	0.001
CARDINAL HEALTH - CO	392-03	12	22	-	-	-	-	-	-	-	-	-	-	-	34	22	0.229	0.010
CARDINAL HEALTH - CT	00163	23	10	-	-	-	-	-	-	-	-	-	-	-	33	10	0.083	0.008
CARDINAL HEALTH - FL	1264-9	12	9	-	-	-	-	-	-	-	-	-	-	-	21	9	0.084	0.009
CARDINAL HEALTH - FL	3453-1	1	8	-	-	-	-	-	-	-	-	-	-	-	9	8	0.166	0.027
CARDINAL HEALTH - FL	3453-2	19	6	1	-	-	-	-	-	-	-	-	-	-	26	7	0.189	0.027
CARDINAL HEALTH - FL	3453-3	7	13	-	-	-	-	-	-	-	-	-	-	-	20	13	0.118	0.009
CARDINAL HEALTH - FL	3453-5	6	11	-	-	-	-	-	-	-	-	-	-	-	17	11	0.128	0.012
CARDINAL HEALTH - FL	3453-6	11	7	-	-	-	-	-	-	-	-	-	-	-	18	7	0.036	0.005
CARDINAL HEALTH - FL	3453-7	12	5	-	-	-	-	-	-	-	-	-	-	-	17	5	0.046	0.009
CARDINAL HEALTH - FL	3453-8	8	17	-	-	-	-	-	-	-	-	-	-	-	25	17	0.247	0.015

Number of Individuals with Whole Body Doses in the Ranges (rems)*

APPENDIX A

Table A - Annual TEDE for Agreement State Licensees

2002 (continued)

MANUFACTURING AND DISTRIBUTION – NUCLEAR PHARMACIES – 02500

PROGRAM CODE - LICENSEE NAME	LICENSE #	No Meas. Exposure	Number of Individuals with Whole Body Doses in the Ranges (rems)*												Total Number Monitored	Number with Meas. Dose	Total Collective TEDE (person-rem)	Average Meas. TEDE (rem)
			Meas. <0.10	0.10-0.25	0.25-0.50	0.50-0.75	0.75-1.00	1.00-2.00	2.00-3.00	3.00-4.00	4.00-5.00	5.00-6.00	6.00-12.00	>12.0				
CARDINAL HEALTH - FL	3453-9	5	10	-	-	-	-	-	-	-	-	-	-	-	15	10	0.134	0.013
CARDINAL HEALTH - GA	GA-467-1MD	3	19	-	-	-	-	-	-	-	-	-	-	-	22	19	0.220	0.012
CARDINAL HEALTH - GA	GA-467-2MD	8	5	-	-	-	-	-	-	-	-	-	-	-	13	5	0.032	0.006
CARDINAL HEALTH - GA	GA-467-3MD	5	-	-	-	-	-	-	-	-	-	-	-	-	5	-	-	-
CARDINAL HEALTH - GA	GA-823-2MD	9	5	-	-	-	-	-	-	-	-	-	-	-	14	5	0.053	0.011
CARDINAL HEALTH - IA	0043-1-77-NP	9	6	-	-	-	-	-	-	-	-	-	-	-	15	6	0.021	0.004
CARDINAL HEALTH - IL	IL-01721-01	22	32	-	-	-	-	-	-	-	-	-	-	-	54	32	0.278	0.009
CARDINAL HEALTH - IN	R0358-45	11	17	-	-	-	-	-	-	-	-	-	-	-	28	17	0.212	0.012
CARDINAL HEALTH - KS	20-C495-01	6	3	-	-	-	-	-	-	-	-	-	-	-	9	3	0.022	0.007
CARDINAL HEALTH - KY	202-204-32	16	3	-	-	-	-	-	-	-	-	-	-	-	19	3	0.035	0.012
CARDINAL HEALTH - KY	202-206-32	1	19	-	-	-	-	-	-	-	-	-	-	-	20	19	0.231	0.012
CARDINAL HEALTH - LA	LA-3385-01	15	11	-	-	-	-	-	-	-	-	-	-	-	26	11	0.121	0.011
CARDINAL HEALTH - LA	LA-7096-01	-	1	-	-	-	-	-	-	-	-	-	-	-	1	1	0.032	0.032
CARDINAL HEALTH - MA	4008	-	1	1	1	1	-	-	-	-	-	-	-	-	4	4	1.084	0.271
CARDINAL HEALTH - MA	42-0146	11	23	-	-	1	-	-	-	-	-	-	-	-	35	24	0.774	0.032
CARDINAL HEALTH - MD	MD-05-058-01	12	13	-	-	-	-	-	-	-	-	-	-	-	25	13	0.121	0.009
CARDINAL HEALTH - MD	MD-33-158-01	7	19	-	-	-	-	-	-	-	-	-	-	-	26	19	0.183	0.010
CARDINAL HEALTH - MI	1549-4	15	20	-	-	-	-	-	-	-	-	-	-	-	35	20	0.243	0.012
CARDINAL HEALTH - MN	1037-205-89	30	30	-	-	-	-	-	-	-	-	-	-	-	60	30	0.514	0.017
CARDINAL HEALTH - MO	2840	1	21	-	-	-	-	-	-	-	-	-	-	-	22	21	0.494	0.024
CARDINAL HEALTH - MO	6141	-	1	2	1	-	-	-	-	-	-	-	-	-	4	4	0.736	0.184
CARDINAL HEALTH - MS	MS-493-01	6	3	-	-	-	-	-	-	-	-	-	-	-	9	3	0.042	0.014
CARDINAL HEALTH - NC	041-0794-3	4	3	-	-	-	-	-	-	-	-	-	-	-	7	3	0.006	0.002
CARDINAL HEALTH - NC	060-0794-1	13	13	-	-	-	-	-	-	-	-	-	-	-	26	13	0.173	0.013
CARDINAL HEALTH - NC	092-1175-1	3	6	-	-	-	-	-	-	-	-	-	-	-	9	6	0.117	0.020
CARDINAL HEALTH - NE	01-65-01	12	17	-	-	-	-	-	-	-	-	-	-	-	29	17	0.117	0.007
CARDINAL HEALTH - NM	RP396-02	5	6	-	-	-	-	-	-	-	-	-	-	-	11	6	0.036	0.006
CARDINAL HEALTH - NV	03-11-0150-01	2	16	-	-	-	-	-	-	-	-	-	-	-	18	16	0.206	0.013
CARDINAL HEALTH - NY	2306-3119MD	17	19	-	-	-	-	-	-	-	-	-	-	-	36	19	0.318	0.017
CARDINAL HEALTH - NY	C2328	16	8	-	-	-	-	-	-	-	-	-	-	-	24	8	0.066	0.008
CARDINAL HEALTH - NY	C2364	10	7	-	-	-	-	-	-	-	-	-	-	-	17	7	0.060	0.009
CARDINAL HEALTH - NY	C2449	10	8	-	-	-	-	-	-	-	-	-	-	-	18	8	0.051	0.006
CARDINAL HEALTH - NY	C2588	9	5	-	-	-	-	-	-	-	-	-	-	-	14	5	0.027	0.005

APPENDIX A

Table A - Annual TEDE for Agreement State Licensees

2002 (continued)

PROGRAM CODE - LICENSEE NAME	LICENSE#	No. Meas. Exposure	Meas. <0.10	Number of Individuals with Whole Body Doses in the Ranges (rems)*											Total Number Monitored	Number with Meas. Dose	Total Collective TEDE (person-rem)	Average Meas. TEDE (rem)
				0.10-0.25	0.25-0.50	0.50-0.75	0.75-1.00	1.00-2.00	2.00-3.00	3.00-4.00	4.00-5.00	5.00-6.00	6.00-12.00	>12.0				
MANUFACTURING AND DISTRIBUTION – NUCLEAR PHARMACIES – 02500																		
CARDINAL HEALTH - NY	C2593	12	11	-	-	-	-	-	-	-	-	-	-	-	23	11	0.080	0.007
CARDINAL HEALTH - NY	C2613	6	6	-	-	-	-	-	-	-	-	-	-	-	12	6	0.045	0.008
CARDINAL HEALTH - OH	02500180000	16	7	-	-	-	-	-	-	-	-	-	-	-	23	7	0.047	0.007
CARDINAL HEALTH - OH	02500250000	37	19	-	-	-	-	-	-	-	-	-	-	-	56	19	0.186	0.010
CARDINAL HEALTH - OH	02500310000	9	3	1	-	-	-	-	-	-	-	-	-	-	13	4	0.227	0.057
CARDINAL HEALTH - OH	02500490001	18	13	1	-	-	-	-	-	-	-	-	-	-	32	14	0.340	0.024
CARDINAL HEALTH - OH	02500580000	12	16	-	-	-	-	-	-	-	-	-	-	-	28	16	0.353	0.022
CARDINAL HEALTH - OH	02500770000	6	3	-	-	-	-	-	-	-	-	-	-	-	9	3	0.016	0.005
CARDINAL HEALTH - OH	02500790000	3	5	-	-	-	-	-	-	-	-	-	-	-	8	5	0.026	0.005
CARDINAL HEALTH - OH	03214310000	-	4	-	-	-	-	-	-	-	-	-	-	-	4	4	0.098	0.025
CARDINAL HEALTH - OH	4853-19	11	5	-	-	-	-	-	-	-	-	-	-	-	16	5	0.155	0.031
CARDINAL HEALTH - OK	OK-19583-02MD	7	3	2	-	-	-	-	-	-	-	-	-	-	12	5	0.350	0.070
CARDINAL HEALTH - OK	OK-23359-02MD	12	7	-	-	-	-	-	-	-	-	-	-	-	19	7	0.037	0.005
CARDINAL HEALTH - OR	ORE-90509	14	13	3	-	-	-	-	-	-	-	-	-	-	30	16	0.679	0.042
CARDINAL HEALTH - PA	PA-0384	10	21	-	-	-	-	-	-	-	-	-	-	-	31	21	0.147	0.007
CARDINAL HEALTH - PA	PA-0385	43	27	1	-	-	-	-	-	-	-	-	-	-	71	28	0.709	0.025
CARDINAL HEALTH - PA	PA-0415	19	12	-	-	-	-	-	-	-	-	-	-	-	31	12	0.118	0.010
CARDINAL HEALTH - PA	PA-0460	6	9	-	-	-	-	-	-	-	-	-	-	-	15	9	0.153	0.017
CARDINAL HEALTH - PA	PA-0643	13	8	-	-	-	-	-	-	-	-	-	-	-	21	8	0.061	0.008
CARDINAL HEALTH - PA	PA-0680	6	13	-	-	-	-	-	-	-	-	-	-	-	19	13	0.149	0.011
CARDINAL HEALTH - RI	3B-114-01	11	12	-	-	-	-	-	-	-	-	-	-	-	23	12	0.062	0.005
CARDINAL HEALTH - SC	448	8	3	-	-	-	-	-	-	-	-	-	-	-	11	3	0.075	0.025
CARDINAL HEALTH - TN	R-19199-B15	8	20	-	-	-	-	-	-	-	-	-	-	-	28	20	0.228	0.011
CARDINAL HEALTH - TN	R-33111-I14	11	6	-	1	-	1	1	-	-	-	-	-	-	20	9	2.855	0.317
CARDINAL HEALTH - TN	R47080-L14	12	13	-	-	-	-	-	-	-	-	-	-	-	25	13	0.079	0.006
CARDINAL HEALTH - TN	R-57025-C15	3	1	-	-	-	-	-	-	-	-	-	-	-	4	1	0.007	0.007
CARDINAL HEALTH - TN	R-71029-B11	6	4	-	-	-	-	-	-	-	-	-	-	-	10	4	0.071	0.018
CARDINAL HEALTH - TN	R-79174-02	1	4	-	-	-	-	-	-	-	-	-	-	-	5	4	0.055	0.014
CARDINAL HEALTH - TN	R-79174-D17	5	5	1	-	-	-	-	-	-	-	-	-	-	11	6	0.129	0.022
CARDINAL HEALTH - TX	L0-1911	26	31	1	-	-	-	-	-	-	-	-	-	-	58	32	0.665	0.021
CARDINAL HEALTH - TX	L0-1999	2	4	-	-	-	-	-	-	-	-	-	-	-	6	4	0.014	0.004
CARDINAL HEALTH - TX	L0-2033	9	19	1	-	-	-	-	-	-	-	-	-	-	29	20	0.435	0.022
CARDINAL HEALTH - TX	L0-2048	13	14	-	-	-	-	-	-	-	-	-	-	-	27	14	0.136	0.010

APPENDIX A

Table A - Annual TEDE for Agreement State Licensees
2002 (continued)

PROGRAM CODE - LICENSEE NAME	LICENSE#	No Meas. Exposure	Meas. <0.10	0.10-0.25	0.25-0.50	0.50-0.75	0.75-1.00	1.00-2.00	2.00-3.00	3.00-4.00	4.00-5.00	5.00-6.00	6.00-12.00	>12.0	Total Number Monitored	Number with Meas. Dose	Total Collective TEDE (person-rem)	Average Meas. TEDE (rem)
MANUFACTURING AND DISTRIBUTION – NUCLEAR PHARMACIES – 02500																		
CARDINAL HEALTH - TX	LO-2117	11	9	-	-	-	-	-	-	-	-	-	-	-	20	9	0.114	0.013
CARDINAL HEALTH - TX	LO-2737	2	3	-	-	-	-	-	-	-	-	-	-	-	5	3	0.033	0.011
CARDINAL HEALTH - TX	LO-2905	8	1	-	-	-	-	-	-	-	-	-	-	-	9	1	0.003	0.003
CARDINAL HEALTH - TX	LO-2587	6	-	-	-	-	-	-	-	-	-	-	-	-	6	-	-	-
CARDINAL HEALTH - TX	LO-3398	3	4	-	-	-	-	-	-	-	-	-	-	-	7	4	0.022	0.006
CARDINAL HEALTH - TX	LO-4043	7	1	-	-	-	-	-	-	-	-	-	-	-	8	1	0.003	0.003
CARDINAL HEALTH - TX	LO-4573	7	2	-	-	-	-	-	-	-	-	-	-	-	9	2	0.010	0.005
CARDINAL HEALTH - TX	LO-4781	7	12	-	-	-	-	-	-	-	-	-	-	-	19	12	0.279	0.023
CARDINAL HEALTH - TX	LO-5536	-	3	1	-	-	-	-	-	-	-	-	-	-	4	4	0.347	0.087
CARDINAL HEALTH - US	074-1180-1	2	3	-	-	-	-	-	-	-	-	-	-	-	5	3	0.020	0.007
CARDINAL HEALTH - US	CH-US-001	47	139	8	1	-	-	-	-	1	-	-	-	-	196	149	9.003	0.060
CARDINAL HEALTH - VA	760-34-1	9	25	-	-	-	-	-	-	-	-	-	-	-	34	25	0.149	0.006
CARDINAL HEALTH - WA	WN-NP003-1	14	10	-	-	-	-	-	-	-	-	-	-	-	24	10	0.077	0.008
CARDINAL HEALTH - WA	WN-NP004-1	9	5	-	-	-	-	-	-	-	-	-	-	-	14	5	0.042	0.008
CARDINAL HEALTH - WA	WN-NF011-1	1	8	-	-	-	-	-	-	-	-	-	-	-	9	8	0.098	0.012
CARDINAL HEALTH - WI	087-1312-01	8	5	-	-	-	-	-	-	-	-	-	-	-	13	5	0.020	0.004
CARDINAL HEALTH - WI	141-1306-01	2	5	-	1	-	-	-	-	-	-	-	-	-	8	6	0.389	0.065
GE HEALTHCARE - ANAHEIM	4810-30	11	22	-	-	-	-	-	-	-	-	-	-	-	33	22	0.538	0.024
GE HEALTHCARE - SACRAMENTO	4809-34	14	7	4	-	-	-	-	-	-	-	-	-	-	25	11	0.821	0.075
GE HEALTHCARE - SAN DIEGO	5796-37	21	2	-	-	-	-	-	-	-	-	-	-	-	23	2	0.032	0.016
GE HEALTHCARE - SAN JOSE	4811-43	6	5	3	-	-	1	-	-	-	-	-	-	-	15	9	0.861	0.096
GE HEALTHCARE - VAN NUYS	5143-19	17	2	2	-	-	-	-	-	-	-	-	-	-	21	4	0.464	0.116
MORAVEK BIOCHEMICALS, INC.	2960-30	-	3	11	5	1	-	-	-	-	-	-	-	-	20	20	4.295	0.215
WYLE LABORATORIES	FL-2953-1	18	4	1	-	-	-	-	-	-	-	-	-	-	23	5	0.195	0.039
Total	123	1,134	1,334	51	13	2	1	1	-	1	-	-	-	-	2,537	1,403	40.065	0.029
WELL LOGGING BYPRODUCT AND/OR SNM SEALED SOURCES ONLY - 03111																		
ISOTOPE PRODUCTS LABS	1509-19	3	27	7	11	2	6	13	-	-	-	-	-	-	69	66	32.331	0.490
Total	1	3	27	7	11	2	6	13	-	-	-	-	-	-	69	66	32.331	0.490
MEASURING SYSTEMS PORTABLE GAUGES - 03121																		
CPN INTERNATIONAL, INC.	1100-07	1	4	3	2	1	1	-	-	-	-	-	-	-	12	11	2.795	0.254
Total	1	1	4	3	2	1	1	-	-	-	-	-	-	-	12	11	2.795	0.254

Number of Individuals with Whole Body Doses in the Ranges (rems)*

APPENDIX A

Table A - Annual TEDE for Agreement State Licensees

2002 (continued)

PROGRAM CODE - LICENSEE NAME	LICENSE#	No Meas. Exposure	Meas. <0.10	0.10-0.25	0.25-0.50	0.50-0.75	0.75-1.00	1.00-2.00	2.00-3.00	3.00-4.00	4.00-5.00	5.00-6.00	6.00-12.00	>12.0	Total Number Monitored	Number with Meas. Dose	Total Collective TEDE (person-rem)	Average Meas. TEDE (rem)
MANUFACTURING AND DISTRIBUTION – OTHER - 03214																	
HOCHIKI AMERICA CORP	2090-30	18	1	-	-	-	-	-	-	-	-	-	-	-	19	1	0.082	0.082
J. L. SHEPHERD AND ASSOCIATES	CA 1777-19	9	13	2	-	2	1	-	-	-	-	-	-	-	27	18	2.775	0.154
Total	2	27	14	2	-	2	1	-	-	-	-	-	-	-	46	19	2.857	0.150
WASTE DISPOSAL SERVICE PROCESSING AND/OR REPACKAGING - 03234																	
ENVIRONMENTAL MANAGEMENT & CONTROLS	3546-50	1	5	3	2	-	-	-	-	-	-	-	-	-	11	10	1.171	0.117
THOMAS GRAY & ASSOCIATES, INC.	CA-2105-30	5	2	1	-	-	-	-	-	-	-	-	-	-	8	3	0.133	0.044
Total	2	6	7	4	2	-	-	-	-	-	-	-	-	-	19	13	1.304	0.100
INDUSTRIAL RADIOGRAPHY – FIXED LOCATION – 03310																	
METALTEK - WISCONSIN CENTRIFUGAL DIVISION	133-1181-01	3	4	2	-	2	-	-	-	-	-	-	-	-	11	8	1.783	0.223
SONGS - RADIOGRAPHY GROUP	CA-5244-30	12	-	-	-	-	-	-	-	-	-	-	-	-	12	-	-	-
Total	2	15	4	2	-	2	-	-	-	-	-	-	-	-	23	8	1.783	0.223
INDUSTRIAL RADIOGRAPHY – TEMPORARY JOB SITE – 03320																	
ACUREN INSPECTION, INC	997-01	1	2	1	4	1	2	1	-	-	-	-	-	-	12	11	5.637	0.512
ALPHA-OMEGA SERVICES, INC.	3925-19	1	12	5	2	1	-	-	-	-	-	-	-	-	21	20	2.501	0.125
ANVIL INTERNATIONAL, INC.	RI-3D-064-01	-	5	1	-	-	-	-	-	-	-	-	-	-	6	6	0.369	0.062
CONSTRUCTION MATERIALS TESTING, INC.	0799-07	1	1	1	2	1	1	-	-	-	-	-	-	-	7	6	2.860	0.477
CONTRA COSTA INSPECTION	CA-1885-07	-	2	-	2	-	-	-	-	-	-	-	-	-	4	4	1.065	0.266
INTERNATIONAL INSPECTION, INC.	5046-19	4	2	-	1	2	2	-	-	-	-	-	-	-	11	7	3.240	0.463
NDE SERVICES, INC	406-01	1	2	1	4	1	-	-	1	-	-	-	-	-	10	9	4.745	0.527
NDT LABORATORIES, INC.	1651-43	-	1	2	-	-	-	-	-	-	-	-	-	-	3	3	0.320	0.107
TC INSPECTION, INC.	5299-07	12	14	4	2	3	1	1	-	-	-	-	-	-	37	25	5.586	0.223
TESTING ENGINEERS, INC.	1898-01	-	2	2	2	1	-	-	-	-	-	-	-	-	7	7	1.945	0.278
WESTERN INDUSTRIAL X-RAY	4424-48	1	3	1	2	1	-	-	-	-	-	-	-	-	8	7	2.577	0.368
WESTERN X-RAY CORPORATION	5324-56	-	-	5	3	1	-	-	-	-	-	-	-	-	9	9	2.470	0.274
Total	12	21	46	22	23	11	8	3	1	-	-	-	-	-	135	114	33.315	0.292
IRRADIATORS - OTHER - LESS THAN 10000 CURIES - 03511																	
INDUSTRIAL NUCLEAR CO., INC.	2229-01	6	2	1	1	1	-	1	-	-	-	-	-	-	12	6	3.190	0.532
Total	1	6	2	1	1	1	-	1	-	-	-	-	-	-	12	6	3.190	0.532

APPENDIX A

Table A - Annual TEDE for Agreement State Licensees
2003

PROGRAM CODE - LICENSEE NAME	LICENSE#	No Meas. Exposure	Meas. <0.10	0.10-0.25	0.25-0.50	0.50-0.75	0.75-1.00	1.00-2.00	2.00-3.00	3.00-4.00	4.00-5.00	5.00-6.00	6.00-12.00	>12.0	Total Number Monitored	Number with Meas. Dose	Total Collective TEDE (person-rem)	Average Meas. TEDE (rem)
MANUFACTURING AND DISTRIBUTION – NUCLEAR PHARMACIES – 02500																		
CARDINAL HEALTH - AK	0707RA907-912	4	2	-	-	-	-	-	-	-	-	-	-	-	6	2	0.006	0.003
CARDINAL HEALTH - AL	1068	4	9	1	-	-	-	-	-	-	-	-	-	-	14	10	0.370	0.037
CARDINAL HEALTH - AL	1168	4	22	12	9	4	2	3	-	-	-	-	-	-	56	52	14.459	0.278
CARDINAL HEALTH - AR	ARK-6-2-02500	10	37	16	5	3	-	3	-	-	-	-	-	-	74	64	10.998	0.172
CARDINAL HEALTH - AZ	01-084	-	6	1	1	1	-	-	-	-	-	-	-	-	9	9	1.168	0.130
CARDINAL HEALTH - AZ	07-123	5	38	9	4	-	-	-	-	-	-	-	-	-	56	51	3.638	0.071
CARDINAL HEALTH - CA	2891-37	3	11	1	3	-	-	-	-	-	-	-	-	-	18	15	1.428	0.095
CARDINAL HEALTH - CA	3317-19	1	14	-	-	-	-	-	-	-	-	-	-	-	15	14	0.265	0.019
CARDINAL HEALTH - CA	3426-45	4	7	1	-	-	-	-	-	-	-	-	-	-	12	8	0.237	0.030
CARDINAL HEALTH - CA	3469-1	-	-	3	1	-	-	1	-	-	-	-	-	-	5	5	2.188	0.438
CARDINAL HEALTH - CA	3469-2	2	1	2	1	2	-	1	-	-	-	-	-	-	9	7	2.954	0.422
CARDINAL HEALTH - CA	3673-34	10	31	2	-	-	-	-	-	-	-	-	-	-	43	33	1.009	0.031
CARDINAL HEALTH - CA	3822-19	7	30	11	5	1	-	-	-	-	-	-	-	-	54	47	5.030	0.107
CARDINAL HEALTH - CA	3832-01	-	17	5	6	-	-	-	-	-	-	-	-	-	28	28	3.554	0.127
CARDINAL HEALTH - CA	4905-10	5	11	4	1	-	-	-	-	-	-	-	-	-	21	16	1.393	0.087
CARDINAL HEALTH - CA	4999-30	1	25	7	1	-	-	-	-	-	-	-	-	-	34	33	2.481	0.075
CARDINAL HEALTH - CA	5218-36	4	10	4	3	-	-	1	-	-	-	-	-	-	22	18	3.577	0.199
CARDINAL HEALTH - CA	5905-15	-	6	-	-	-	-	-	-	-	-	-	-	-	6	6	0.130	0.022
CARDINAL HEALTH - CA	5910-50	-	4	1	-	-	-	-	-	-	-	-	-	-	5	5	0.206	0.041
CARDINAL HEALTH - CA	6321-45	-	4	-	-	-	-	-	-	-	-	-	-	-	4	4	0.014	0.004
CARDINAL HEALTH - CA	6691-33	2	11	-	-	-	-	-	-	-	-	-	-	-	13	11	0.272	0.025
CARDINAL HEALTH - CA	6924-36	-	-	-	2	-	1	-	1	-	-	-	-	-	4	4	4.777	**1.194**
CARDINAL HEALTH - CA	6925-19	7	7	-	-	-	-	-	-	-	-	-	-	-	14	7	0.255	0.036
CARDINAL HEALTH - CO	392-01	9	33	3	5	-	-	-	-	-	-	-	-	-	50	41	2.668	0.065
CARDINAL HEALTH - CO	392-03	6	33	2	5	-	-	-	-	-	-	-	-	-	46	40	2.602	0.065
CARDINAL HEALTH - CT	00163	2	32	5	-	1	-	-	-	-	-	-	-	-	40	38	1.886	0.050
CARDINAL HEALTH - FL	1264-9	2	15	4	1	-	-	-	-	-	-	-	-	-	22	20	1.145	0.057
CARDINAL HEALTH - FL	3273-7	-	10	1	3	1	1	-	-	-	-	-	-	-	16	16	3.534	0.221
CARDINAL HEALTH - FL	3453-1	4	10	1	1	-	-	-	-	-	-	-	-	-	16	12	0.660	0.055
CARDINAL HEALTH - FL	3453-2	5	26	5	-	-	-	-	-	-	-	-	-	-	36	31	1.364	0.044
CARDINAL HEALTH - FL	3453-3	5	12	7	-	-	-	-	-	-	-	-	-	-	24	19	1.560	0.082
CARDINAL HEALTH - FL	3453-5	-	15	3	3	-	-	-	-	-	-	-	-	-	21	21	2.159	0.103

APPENDIX A
Table A - Annual TEDE for Agreement State Licensees
2003 (continued)

MANUFACTURING AND DISTRIBUTION – NUCLEAR PHARMACIES – 02500

PROGRAM CODE - LICENSEE NAME	LICENSE#	No Meas. Exposure	Meas. <0.10	Number of Individuals with Whole Body Doses in the Ranges (rems)*											Total Number Monitored	Number with Meas. Dose	Total Collective TEDE (person-rem)	Average Meas. TEDE (rem)
				0.10-0.25	0.25-0.50	0.50-0.75	0.75-1.00	1.00-2.00	2.00-3.00	3.00-4.00	4.00-5.00	5.00-6.00	6.00-12.00	>12.0				
CARDINAL HEALTH - FL	3453-6	4	11	4	-	-	-	-	-	-	-	-	-	-	19	15	1.093	0.073
CARDINAL HEALTH - FL	3453-7	2	16	2	2	-	-	-	-	-	-	-	-	-	22	20	1.722	0.086
CARDINAL HEALTH - FL	3453-8	6	18	5	1	-	-	-	-	-	-	-	-	-	30	24	1.597	0.067
CARDINAL HEALTH - FL	3453-9	2	14	2	3	-	-	-	-	-	-	-	-	-	21	19	1.648	0.087
CARDINAL HEALTH - FL	FL-3273-1	2	-	-	-	-	-	-	-	-	-	-	-	-	2	-	-	-
CARDINAL HEALTH - FL	FL-3273-3	-	4	-	-	-	-	-	-	-	-	-	-	-	4	4	0.153	0.038
CARDINAL HEALTH - FL	FL-3273-6	2	10	-	-	-	-	-	-	-	-	-	-	-	12	10	0.374	0.037
CARDINAL HEALTH - GA	GA-467-1MD	1	15	6	3	-	-	-	-	-	-	-	-	-	25	24	2.512	0.105
CARDINAL HEALTH - GA	GA-467-2MD	7	9	-	-	-	-	-	-	-	-	-	-	-	16	9	0.228	0.025
CARDINAL HEALTH - GA	GA-467-3MD	2	3	-	-	-	-	-	-	-	-	-	-	-	5	3	0.015	0.005
CARDINAL HEALTH - GA	GA-823-2MD	2	16	9	-	-	-	-	-	-	-	-	-	-	27	25	2.084	0.083
CARDINAL HEALTH - IA	0043-1-77-NP	4	14	1	-	-	-	-	-	-	-	-	-	-	19	15	0.550	0.037
CARDINAL HEALTH - IL	IL-01721-01	5	36	15	3	-	-	-	-	-	-	-	-	-	59	54	4.947	0.092
CARDINAL HEALTH - IN	R0358-45	6	28	3	-	-	-	-	-	-	-	-	-	-	37	31	1.181	0.038
CARDINAL HEALTH - KS	20-C495-01	5	15	4	-	-	-	-	-	-	-	-	-	-	24	19	1.140	0.060
CARDINAL HEALTH - KY	202-204-32	10	18	7	1	1	-	-	-	-	-	-	-	-	37	27	2.215	0.082
CARDINAL HEALTH - KY	202-206-32	3	21	10	1	-	-	-	-	-	-	-	-	-	35	32	2.993	0.094
CARDINAL HEALTH - LA	LA-10217-L01	-	5	-	2	-	-	-	-	-	-	-	-	-	7	7	0.847	0.121
CARDINAL HEALTH - LA	LA-10336-L01	4	1	-	-	1	1	-	-	-	-	-	-	-	7	3	0.440	0.147
CARDINAL HEALTH - LA	LA-3385-L01	13	23	6	1	1	1	-	-	-	-	-	-	-	45	32	2.970	0.093
CARDINAL HEALTH - LA	LA-5115-L01	9	11	1	-	-	-	-	-	-	-	-	-	-	21	12	0.399	0.033
CARDINAL HEALTH - LA	LA-5119-L01	1	7	1	1	1	-	1	-	1	-	-	-	-	13	12	5.849	0.487
CARDINAL HEALTH - LA	LA-5394-L01	2	9	3	2	-	1	2	-	-	1	-	-	-	20	18	11.428	0.938
CARDINAL HEALTH - LA	LA-7096-L01	2	8	2	-	-	-	-	-	-	-	-	-	-	12	10	0.579	0.058
CARDINAL HEALTH - MA	4008	-	2	-	1	1	1	-	-	-	1	-	-	-	6	6	6.163	1.027
CARDINAL HEALTH - MA	42-0146	7	26	14	3	3	-	3	-	-	-	-	-	-	56	49	10.418	0.213
CARDINAL HEALTH - MD	MD-05-058-01	10	17	5	-	-	-	-	-	-	-	-	-	-	32	22	1.231	0.056
CARDINAL HEALTH - MD	MD-33-198-01	8	38	7	-	-	-	-	-	-	-	-	-	-	53	45	2.024	0.045
CARDINAL HEALTH - MI	1549-4	8	30	5	4	-	-	-	-	-	-	-	-	-	47	39	2.734	0.070
CARDINAL HEALTH - MN	1037-205-89	8	46	8	5	-	-	-	-	-	-	-	-	-	67	59	4.032	0.068
CARDINAL HEALTH - MO	2840	2	27	5	1	-	-	-	-	-	-	-	-	-	35	33	1.590	0.048
CARDINAL HEALTH - MO	6141	-	-	-	-	1	-	3	-	-	-	-	-	-	4	4	4.733	1.183
CARDINAL HEALTH - MS	MS-493-01	7	9	-	5	-	-	-	-	-	-	-	-	-	21	14	1.855	0.133

APPENDIX A

Table A - Annual TEDE for Agreement State Licensees
2003 (continued)

MANUFACTURING AND DISTRIBUTION – NUCLEAR PHARMACIES – 02500

PROGRAM CODE - LICENSEE NAME	LICENSE#	No Meas. Exposure	Meas. <0.10	Number of Individuals with Whole Body Doses in the Ranges (rems)												Total Number Monitored	Number with Meas. Dose	Total Collective TEDE (person-rem)	Average Meas. TEDE (rem)
				0.10-0.25	0.25-0.50	0.50-0.75	0.75-1.00	1.00-2.00	2.00-3.00	3.00-4.00	4.00-5.00	5.00-6.00	6.00-12.00	>12.0				
CARDINAL HEALTH - MS	MS-781-01	-	4	-	3	-	-	1	1	1	-	-	-	-	10	10	7.961	0.796
CARDINAL HEALTH - MS	MS-915-01	3	1	-	-	-	-	-	-	-	-	-	-	-	4	1	0.010	0.010
CARDINAL HEALTH - MS	MS-924-01	4	3	1	1	-	-	-	-	-	-	-	-	-	9	5	0.680	0.136
CARDINAL HEALTH - NC	041-3794-3	-	8	-	-	-	-	-	-	-	-	-	-	-	8	8	0.259	0.032
CARDINAL HEALTH - NC	060-3794-1	7	20	2	-	-	-	-	-	-	-	-	-	-	29	22	0.775	0.035
CARDINAL HEALTH - NC	092-1175-1	-	7	3	2	-	-	-	-	-	-	-	-	-	12	12	1.464	0.122
CARDINAL HEALTH - NE	01-65-01	2	34	5	2	-	-	-	-	-	-	-	-	-	43	41	2.400	0.059
CARDINAL HEALTH - NM	RP356-02	1	10	1	-	-	-	-	-	-	-	-	-	-	12	11	0.396	0.036
CARDINAL HEALTH - NV	03-11-0150-01	1	5	6	-	-	-	-	-	-	-	-	-	-	12	11	1.121	0.102
CARDINAL HEALTH - NV	092-C794-6	2	7	2	2	-	-	-	-	-	-	-	-	-	13	11	1.295	0.118
CARDINAL HEALTH - NY	2306-3119MD	10	28	4	1	-	-	-	-	-	-	-	-	-	43	33	1.844	0.056
CARDINAL HEALTH - NY	C2328	6	15	3	1	-	-	-	-	-	-	-	-	-	25	19	0.979	0.052
CARDINAL HEALTH - NY	C2364	10	20	-	-	-	-	-	-	-	-	-	-	-	30	20	0.302	0.015
CARDINAL HEALTH - NY	C2449	4	16	2	-	-	-	-	-	-	-	-	-	-	22	18	0.606	0.034
CARDINAL HEALTH - NY	C2586	6	16	1	-	-	-	-	-	-	-	-	-	-	23	17	0.510	0.030
CARDINAL HEALTH - NY	C2593	8	21	-	-	-	1	-	-	-	-	-	-	-	30	22	0.857	0.039
CARDINAL HEALTH - NY	C2613	3	13	1	-	1	-	-	-	-	-	-	-	-	18	15	1.244	0.083
CARDINAL HEALTH - OH	02500180000	12	15	2	-	-	-	-	-	-	-	-	-	-	29	17	0.659	0.039
CARDINAL HEALTH - OH	02500250000	14	40	8	2	-	-	-	-	-	-	-	-	-	64	50	2.590	0.052
CARDINAL HEALTH - OH	02500310000	3	8	-	1	1	-	-	-	-	-	-	-	-	13	10	1.077	0.108
CARDINAL HEALTH - OH	02500490001	14	22	5	-	-	-	-	-	-	-	-	-	-	41	27	1.126	0.042
CARDINAL HEALTH - OH	02500530000	4	18	4	2	-	-	-	-	-	-	-	-	-	28	24	1.760	0.073
CARDINAL HEALTH - OH	02500770000	3	8	-	-	-	-	-	-	-	-	-	-	-	11	8	0.267	0.033
CARDINAL HEALTH - OH	02500790000	1	9	1	-	-	-	-	-	-	-	-	-	-	11	10	0.391	0.039
CARDINAL HEALTH - OH	03214430000	-	1	2	-	1	1	-	-	-	-	-	-	-	5	5	1.802	0.360
CARDINAL HEALTH - OH	4853-19	8	19	1	-	-	-	-	-	-	-	-	-	-	28	20	0.628	0.031
CARDINAL HEALTH - OK	OK-19583-02MD	3	15	7	4	-	-	-	-	-	-	-	-	-	29	26	3.020	0.116
CARDINAL HEALTH - OK	OK-23359-02MD	4	17	1	-	-	-	-	-	-	-	-	-	-	22	18	0.679	0.038
CARDINAL HEALTH - OR	ORE-90509	-	8	3	2	-	-	-	-	-	-	-	-	-	13	13	1.278	0.098
CARDINAL HEALTH - OR	ORE-90703	-	8	3	3	1	1	4	-	-	-	-	-	-	20	20	10.233	0.512
CARDINAL HEALTH - OR	ORE-90514	8	15	4	1	1	1	-	-	-	-	-	-	-	30	22	2.749	0.125
CARDINAL HEALTH - PA	PA-0384	1	32	8	1	-	-	-	-	-	-	-	-	-	42	41	2.309	0.056
CARDINAL HEALTH - PA	PA-0385	9	53	11	11	3	-	-	-	-	-	-	-	-	87	78	9.201	0.118

APPENDIX A
Table A - Annual TEDE for Agreement State Licensees
2003 (continued)

MANUFACTURING AND DISTRIBUTION – NUCLEAR PHARMACIES – 02500

PROGRAM CODE - LICENSEE NAME	LICENSE#	No Meas. Exposure	Meas. <0.10	0.10-0.25	0.25-0.50	0.50-0.75	0.75-1.00	1.00-2.00	2.00-3.00	3.00-4.00	4.00-5.00	5.00-6.00	6.00-12.00	>12.0	Total Number Monitored	Number with Meas. Dose	Total Collective TEDE (person-rem)	Average Meas. TEDE (rem)
CARDINAL HEALTH - PA	PA-0415	-	26	7	1										34	34	2.251	0.066
CARDINAL HEALTH - PA	PA-0460	1	16												17	16	0.374	0.023
CARDINAL HEALTH - PA	PA-0643	4	19	2											25	21	0.787	0.037
CARDINAL HEALTH - PA	PA-0680	2	19	1											22	20	0.723	0.036
CARDINAL HEALTH - RI	3B-114-01	4	25	1											30	26	0.869	0.033
CARDINAL HEALTH - SC	448	2	9	1	2										14	12	0.859	0.072
CARDINAL HEALTH - SD	47111	-	12	2		1									15	15	1.465	0.098
CARDINAL HEALTH - TN	R-19199-B15	2	26	16	1										45	43	3.500	0.081
CARDINAL HEALTH - TN	R-33111-I14	3	18	2											23	20	0.875	0.044
CARDINAL HEALTH - TN	R47080-L14	3	19	3											25	22	0.893	0.041
CARDINAL HEALTH - TN	R-57025-C15	1	6			1									8	7	0.635	0.091
CARDINAL HEALTH - TN	R-71029-B11	3	7	1	1										12	9	0.753	0.084
CARDINAL HEALTH - TN	R-79174-02	-	1	2		1									4	4	1.011	0.253
CARDINAL HEALTH - TN	R-79174-D17	1	10	1	1	1									14	13	1.285	0.099
CARDINAL HEALTH - TX	L0-1911	18	41	8	3	1									71	53	3.916	0.074
CARDINAL HEALTH - TX	L0-1999	-	6												6	6	0.072	0.012
CARDINAL HEALTH - TX	L0-2033	2	24	5	1	1	1								34	32	3.356	0.105
CARDINAL HEALTH - TX	L0-2048	11	24	5											40	29	1.399	0.048
CARDINAL HEALTH - TX	L0-2117	3	17	2	1										23	20	0.963	0.048
CARDINAL HEALTH - TX	L0-2737	1	4												5	4	0.089	0.022
CARDINAL HEALTH - TX	L0-2905	5	7	1											13	8	0.153	0.019
CARDINAL HEALTH - TX	L0-2987	2	-												2	-	-	-
CARDINAL HEALTH - TX	L0-3398	1	11												12	11	0.158	0.014
CARDINAL HEALTH - TX	L0-4043	-	8												8	8	0.206	0.026
CARDINAL HEALTH - TX	L0-4573	1	19	1											21	20	0.501	0.025
CARDINAL HEALTH - TX	L0-4781	10	10	2	2	1									23	13	1.192	0.092
CARDINAL HEALTH - TX	L0-4785	-	9	1	1										11	11	0.993	0.090
CARDINAL HEALTH - TX	L0-5536	1	1	1		1		1							5	5	2.428	0.486
CARDINAL HEALTH - US	074-1180-1	1	5	1	1										8	7	0.516	0.074
CARDINAL HEALTH - US	CH-US-001	36	96	45	44	8	3	3							235	199	36.061	0.181
CARDINAL HEALTH - UT	UT-1800225	2	12	2	4	1									21	19	2.731	0.144
CARDINAL HEALTH - VA	760-34-1	1	15	28											44	43	4.187	0.097
CARDINAL HEALTH - WA	WN-NP003-1	6	22	3	2										33	27	1.542	0.057
CARDINAL HEALTH - WA	WN-NP004-1	4	21	5	2										32	28	2.155	0.077
CARDINAL HEALTH - WA	WN-NP005-1	3	12	2	2	1									20	17	2.367	0.139

APPENDIX A
Table A - Annual TEDE for Agreement State Licensees
2003 (continued)

PROGRAM CODE - LICENSEE NAME	LICENSE#	No. Meas. Exposure	Meas. <0.10	Number of Individuals with Whole Body Doses in the Ranges (rems)											Total Number Monitored	Number with Meas. Dose	Total Collective TEDE (person-rem)	Average Meas. TEDE (rem)
				0.10-0.25	0.25-0.50	0.50-0.75	0.75-1.00	1.00-2.00	2.00-3.00	3.00-4.00	4.00-5.00	5.00-6.00	6.00-12.00	>12.0				
MANUFACTURING AND DISTRIBUTION – NUCLEAR PHARMACIES – 02500																		
CARDINAL HEALTH - WA	WN-NP011-1	44	24	3	3	-	-	-	-	-	-	-	-	-	74	30	2.471	0.082
CARDINAL HEALTH - WI	025-1-23-01	-	1	-	2	1	1	4	-	-	-	-	-	-	9	9	8.453	0.939
CARDINAL HEALTH - WI	087-1312-01	8	9	2	-	-	-	-	-	-	-	-	-	-	19	11	0.481	0.044
CARDINAL HEALTH - WI	141-1306-01	1	4	2	1	-	-	-	-	-	-	-	-	-	8	7	0.863	0.123
GE HEALTHCARE - ANAHEIM	4810-30	21	13	-	-	-	-	-	-	-	-	-	-	-	34	13	0.529	0.041
GE HEALTHCARE - SACRAMENTO	4809-34	17	3	5	-	-	-	-	-	-	-	-	-	-	25	8	0.775	0.097
GE HEALTHCARE - SAN DIEGO	5796-37	24	6	-	-	-	-	-	-	-	-	-	-	-	30	6	0.129	0.022
GE HEALTHCARE - SAN JOSE	4811-43	3	9	4	2	-	-	-	-	-	-	-	-	-	18	15	1.946	0.130
GE HEALTHCARE - VAN NUYS	5143-13	13	4	2	-	-	-	-	-	-	-	-	-	-	19	6	0.426	0.071
IBA MOLECULAR NORTH AMERICA, INC.	CA-7131-43	3	2	-	-	-	-	-	-	-	-	-	-	-	5	2	0.063	0.032
MALLINCKRODT, INC.	859-01	7	20	1	-	-	-	-	-	-	-	-	-	-	28	21	0.846	0.040
MORAVEK BIOCHEMICALS, INC.	2960-30	-	17	2	-	-	-	1	-	-	-	-	-	-	20	20	2.399	0.120
WYLE LABORATORIES	FL-2953-1	24	3	-	-	-	-	-	-	-	-	-	-	-	27	3	0.060	0.020
Total	**147**	689	2,207	499	223	47	14	31	4	2	2	0	-	-	3,728	3,029	324.649	0.109
WELL LOGGING BYPRODUCT AND/OR SNM SEALED SOURCES ONLY - 03111																		
ISOTOPE PRODUCTS LABS	1509-1E	2	24	8	4	3	7	12	-	-	-	-	-	-	60	58	28.011	0.483
Total	**1**	2	24	8	4	3	7	12	-	-	-	-	-	-	60	58	28.011	0.483
MEASURING SYSTEMS PORTABLE GAUGES - 03121																		
CPN INTERNATIONAL, INC.	1100-07	1	3	4	4	-	-	1	-	-	-	-	-	-	13	12	3.242	0.270
Total	**1**	1	3	4	4	-	-	1	-	-	-	-	-	-	13	12	3.242	0.270
MANUFACTURING AND DISTRIBUTION – OTHER - 03214																		
HOCHIKI AMERICA CORP	2090-30	20	-	-	-	-	-	-	-	-	-	-	-	-	20	-	-	-
J. L. SHEPHERD AND ASSOCIATES	CA 1777-19	12	9	2	2	1	1	-	-	-	-	-	-	-	27	15	3.443	0.230
Total	**2**	32	9	2	2	1	1	-	-	-	-	-	-	-	47	15	3.443	0.230
WASTE DISPOSAL SERVICE PROCESSING AND/OR REPACKAGING - 03234																		
ENVIRONMENTAL MANAGEMENT & CONTROLS	3546-50	1	3	4	-	1	-	-	-	-	-	-	-	-	9	8	1.387	0.173
THOMAS GRAY & ASSOCIATES, INC.	2105-30	6	2	-	-	-	-	-	-	-	-	-	-	-	8	2	0.098	0.049
Total	**2**	7	5	4	-	1	-	-	-	-	-	-	-	-	17	10	1.485	0.149

APPENDIX A

Table A - Annual TEDE for Agreement State Licensees
2003 (continued)

Columns "0.10-0.25" through ">12.0" fall under the spanning header: **Number of Individuals with Whole Body Doses in the Ranges (rems)***

PROGRAM CODE - LICENSEE NAME	LICENSE#	No Meas. Exposure	Meas. <0.10	0.10-0.25	0.25-0.50	0.50-0.75	0.75-1.00	1.00-2.00	2.00-3.00	3.00-4.00	4.00-5.00	5.00-6.00	6.00-12.00	>12.0	Total Number Monitored	Number with Meas. Dose	Total Collective TEDE (person-rem)	Average Meas. TEDE (rem)
INDUSTRIAL RADIOGRAPHY – FIXED LOCATION – 03310																	
LAFAYETTE TESTING SERVICES, INC.	079-1147-01	1	3	-	-	-	-	-	-	-	-	-	-	-	4	3	0.084	0.028
METALTEK - WI CENTRIFUGAL DIVISION	133-1181-01	4	3	1	2	-	-	-	-	-	-	-	-	-	10	6	0.991	0.165
SONGS - RADIOGRAPHY GROUP	CA-5244-30	12	-	-	-	-	-	-	-	-	-	-	-	-	12	-	-	-
WAUKESHA FOUNDRY, INC. - WI	133-1337-01	3	-	-	-	-	-	-	-	-	-	-	-	-	3	-	-	-
Total	**4**	20	6	1	2	-	-	-	-	-	-	-	-	-	29	9	1.075	0.119
INDUSTRIAL RADIOGRAPHY – TEMPORARY JOB SITE – 03320																	
ACTON INSPECTION, LLC.	OK-27438-01	1	2	1	1	1	1	2	-	-	-	-	-	-	9	8	5.237	0.655
ACUREN INSPECTION, INC.	997-01	-	4	1	3	1	-	2	-	-	-	-	-	-	11	11	4.657	0.423
ALPHA-OMEGA SERVICES, INC.	3925-19	1	9	3	2	-	-	-	-	-	-	-	-	-	15	14	1.642	0.117
AMERICAN ENGINEERING TESTING, INC.	073-2012-02	2	7	3	5	-	2	9	2	2	-	-	-	-	32	30	28.982	0.966
ANVIL INTERNATIONAL, INC.	RI-3D064-01	1	2	3	-	-	-	-	-	-	-	-	-	-	6	5	0.496	0.099
COOPER-HEAT-MQS INSPECTION	388-01	1	6	2	2	2	1	2	-	-	-	-	-	-	16	15	6.293	0.420
CONSTRUCTION MATERIALS TESTING, INC	0799-07	1	-	2	2	-	-	-	-	-	-	-	-	-	5	4	0.899	0.225
FTS, INC.	OK-31027-01	-	2	3	2	1	2	1	-	-	-	-	-	-	11	11	5.210	0.474
INTERNATIONAL INSPECTION, INC	5046-19	5	-	2	2	2	-	-	-	-	-	-	-	-	11	6	2.382	0.397
NDE SERVICES, INC.	406-01	2	1	-	4	2	-	-	-	-	-	-	-	-	9	7	2.734	0.391
NDT, INC.	OK-26824-01	-	1	2	-	1	2	1	2	-	-	-	-	-	10	10	13.016	1.302
NDT LABORATORIES, INC.	1651-43	-	1	-	2	-	-	-	-	-	-	-	-	-	3	3	0.680	0.227
NDT SPECIALISTS, INC.	079-1199-01	2	3	3	-	2	-	2	-	-	-	-	-	-	12	10	5.324	0.532
TC INSPECTION, INC.	5299-07	13	6	10	7	6	1	3	-	-	-	-	-	-	46	33	12.739	0.386
TEAM INDUSTRIAL SERVICES, INC.	079-2005-01	1	9	1	-	2	1	-	-	-	-	-	-	-	14	13	2.127	0.164
TESTING ENGINEERS, INC.	1898-01	-	2	1	1	1	1	-	-	-	-	-	-	-	6	6	2.024	0.337
TWIN PORTS TESTING, INC.	031-1317-02	3	-	3	3	2	1	-	-	-	-	-	-	-	12	9	3.617	0.402
WESTERN INDUSTRIAL X-RAY	4424-48	-	-	-	-	1	1	1	-	2	-	-	-	-	5	5	6.078	1.216
WESTERN X-RAY CORP	5324-56	1	1	2	1	-	1	-	-	-	-	-	-	-	6	5	2.231	0.446
WOS TESTING, INC.	035-1358-01	2	6	2	1	-	1	-	-	-	-	-	-	-	12	10	2.362	0.236
YUBA HEAT TRANSFER	OK-13735-01	-	1	1	-	-	-	-	-	-	-	-	-	-	2	2	0.166	0.083
Total	**21**	36	63	43	41	25	10	27	4	4	-	-	-	-	253	217	108.896	0.502
IRRADIATORS - OTHER - LESS THAN 10000 CURIES - 03511																	
INDUSTRIAL NUCLEAR CO., INC.	2229-01	9	-	-	2	-	-	1	1	-	-	-	-	-	13	4	3.989	0.997
Total	**1**	9	-	-	2	-	-	1	1	-	-	-	-	-	13	4	3.989	0.997

APPENDIX A

Table A - Annual TEDE for Agreement State Licensees
2004

MANUFACTURING AND DISTRIBUTION – NUCLEAR PHARMACIES – 02500

PROGRAM CODE - LICENSEE NAME	LICENSE#	No Meas. Exposure	Meas. <0.10	0.10-0.25	0.25-0.50	0.50-0.75	0.75-1.00	1.00-2.00	2.00-3.00	3.00-4.00	4.00-5.00	5.00-6.00	6.00-12.00	>12.0	Total Number Monitored	Number with Meas. Dose	Total Collective TEDE (person-rem)	Average Meas. TEDE (rem)
CARDINAL HEALTH - AK	0707RA907-912	-	6	2	-	-	-	-	-	-	-	-	-	-	8	8	0.518	0.065
CARDINAL HEALTH - AL	1068	2	10	1	-	-	-	-	-	-	-	-	-	-	13	11	0.608	0.055
CARDINAL HEALTH - AL	1168	9	32	15	6	2	1	-	-	-	-	-	-	-	65	56	7.804	0.139
CARDINAL HEALTH - AR	ARK-642-02500	9	47	14	2	-	-	-	-	-	-	-	-	-	72	63	4.421	0.070
CARDINAL HEALTH - AZ	01-064	3	6	-	1	1	-	-	-	-	-	-	-	-	11	8	1.120	0.140
CARDINAL HEALTH - AZ	07-123	8	37	8	3	-	-	-	-	-	-	-	-	-	56	48	3.497	0.073
CARDINAL HEALTH - CA	2891-37	1	12	1	2	1	-	-	-	-	-	-	-	-	17	16	1.880	0.118
CARDINAL HEALTH - CA	3317-19	2	14	3	-	-	-	-	-	-	-	-	-	-	19	17	0.935	0.055
CARDINAL HEALTH - CA	3426-43	-	2	1	3	2	1	-	-	-	-	-	-	-	9	9	3.435	0.382
CARDINAL HEALTH - CA	3469-1	-	-	-	3	-	1	-	-	-	-	-	-	-	4	4	2.216	0.554
CARDINAL HEALTH - CA	3469-2	-	2	2	-	2	-	2	-	-	-	-	-	-	8	8	4.961	0.620
CARDINAL HEALTH - CA	3673-34	12	30	3	-	-	-	-	-	-	-	-	-	-	45	33	0.899	0.027
CARDINAL HEALTH - CA	3822-19	5	36	12	3	1	-	-	-	-	-	-	-	-	57	52	4.679	0.090
CARDINAL HEALTH - CA	3832-C1	-	22	6	3	1	-	-	-	-	-	-	-	-	32	32	3.566	0.111
CARDINAL HEALTH - CA	4905-10	2	19	2	1	-	-	-	-	-	-	-	-	-	24	22	0.991	0.045
CARDINAL HEALTH - CA	4999-33	-	29	8	3	-	-	-	-	-	-	-	-	-	40	40	2.889	0.072
CARDINAL HEALTH - CA	5218-36	2	11	2	2	1	1	1	-	-	-	-	-	-	20	18	3.705	0.206
CARDINAL HEALTH - CA	5905-15	1	5	-	-	-	-	-	-	-	-	-	-	-	6	5	0.148	0.030
CARDINAL HEALTH - CA	5910-50	1	3	1	-	-	-	-	-	-	-	-	-	-	5	4	0.263	0.066
CARDINAL HEALTH - CA	6321-45	1	3	-	-	-	-	-	-	-	-	-	-	-	4	3	0.043	0.014
CARDINAL HEALTH - CA	6691-3E	3	7	-	-	-	-	-	-	-	-	-	-	-	10	7	0.055	0.008
CARDINAL HEALTH - CA	6924-36	-	-	-	-	1	-	1	1	-	-	-	-	-	3	3	4.227	**1.409**
CARDINAL HEALTH - CA	6925-19	1	1	1	2	1	-	1	-	-	-	-	-	-	7	6	3.835	0.639
CARDINAL HEALTH - CO	162-06	1	9	2	2	2	2	-	-	-	-	-	-	-	18	17	3.700	0.218
CARDINAL HEALTH - CO	392-01	24	31	10	4	1	-	-	-	-	-	-	-	-	70	46	4.044	0.088
CARDINAL HEALTH - CO	392-03	6	28	9	1	1	-	-	-	-	-	-	-	-	45	39	2.827	0.072
CARDINAL HEALTH - CT	00163	8	26	9	1	-	-	-	-	-	-	-	-	-	44	36	2.478	0.069
CARDINAL HEALTH - FL	1264-9	3	18	2	1	-	-	-	-	-	-	-	-	-	24	21	0.938	0.045
CARDINAL HEALTH - FL	3273-7	6	14	4	1	-	-	-	-	-	-	-	-	-	25	19	1.130	0.059
CARDINAL HEALTH - FL	3453-1	13	25	3	1	1	-	-	-	-	-	-	-	-	43	30	1.554	0.052
CARDINAL HEALTH - FL	3453-2	7	19	3	-	-	-	-	-	-	-	-	-	-	29	22	0.899	0.041
CARDINAL HEALTH - FL	3453-3	1	21	4	-	-	-	-	-	-	-	-	-	-	26	25	1.240	0.050
CARDINAL HEALTH - FL	3453-5	3	14	4	2	3	-	-	-	-	-	-	-	-	26	23	3.347	0.146

Number of Individuals with Whole Body Doses in the Ranges (rems)*

APPENDIX A

Table A - Annual TEDE for Agreement State Licensees
2004 (continued)

PROGRAM CODE - LICENSEE NAME	LICENSE#	No. Meas. Exposure	Meas. <0.10	Number of Individuals with Whole Body Doses in the Ranges (rems)*											Total Number Monitored	Number with Meas. Dose	Total Collective TEDE (person-rem)	Average Meas. TEDE (rem)
				0.10-0.25	0.25-0.50	0.50-0.75	0.75-1.00	1.00-2.00	2.00-3.00	3.00-4.00	4.00-5.00	5.00-6.00	6.00-12.00	>12.0				
MANUFACTURING AND DISTRIBUTION – NUCLEAR PHARMACIES – 02500																		
CARDINAL HEALTH - FL	3453-6	4	23	5	4	-	-	-	-	-	-	-	-	-	36	32	2.729	0.085
CARDINAL HEALTH - FL	3453-7	4	15	10	3	-	-	-	-	-	-	-	-	-	32	28	3.623	0.129
CARDINAL HEALTH - FL	3453-8	4	19	4	1	-	-	-	-	-	-	-	-	-	28	24	1.900	0.079
CARDINAL HEALTH - FL	3453-9	2	13	2	2	-	-	-	-	-	-	-	-	-	19	17	1.303	0.077
CARDINAL HEALTH - FL	FL-3273-3	-	7	-	-	-	-	-	-	-	-	-	-	-	7	7	0.129	0.018
CARDINAL HEALTH - FL	FL-3273-6	1	10	1	1	-	-	-	-	-	-	-	-	-	13	12	0.622	0.052
CARDINAL HEALTH - GA	GA-467-1MD	1	19	5	4	-	-	-	-	-	-	-	-	-	29	28	3.153	0.113
CARDINAL HEALTH - GA	GA-467-2MD	4	7	1	-	-	-	-	-	-	-	-	-	-	12	8	0.220	0.028
CARDINAL HEALTH - GA	GA-467-3MD	1	4	-	-	-	-	-	-	-	-	-	-	-	5	4	0.047	0.012
CARDINAL HEALTH - GA	GA-823-2MD	1	13	5	2	-	-	-	-	-	-	-	-	-	21	20	2.009	0.100
CARDINAL HEALTH - IA	0043-1-77-NP	3	17	-	-	-	-	-	-	-	-	-	-	-	20	17	0.416	0.024
CARDINAL HEALTH - IL	IL-01721-01	6	44	9	1	-	-	-	-	-	-	-	-	-	60	54	3.122	0.058
CARDINAL HEALTH - IN	R0358-45	8	25	1	1	-	-	-	-	-	-	-	-	-	35	27	0.899	0.033
CARDINAL HEALTH - KS	20-C495-01	11	9	5	3	1	-	-	-	-	-	-	-	-	29	18	3.101	0.172
CARDINAL HEALTH - KY	202-204-32	7	25	4	-	-	-	-	-	-	-	-	-	-	36	29	1.378	0.048
CARDINAL HEALTH - KY	202-206-32	1	20	12	-	-	-	-	-	-	-	-	-	-	33	32	3.040	0.095
CARDINAL HEALTH - LA	LA-10217-L01	-	6	1	-	-	-	-	-	-	-	-	-	-	7	7	0.262	0.037
CARDINAL HEALTH - LA	LA-10336-L01	-	7	1	-	-	-	-	-	-	-	-	-	-	8	8	0.305	0.038
CARDINAL HEALTH - LA	LA-3385-L01	18	18	2	2	1	1	-	-	-	-	-	-	-	42	24	2.944	0.123
CARDINAL HEALTH - LA	LA-5115-L01	2	17	4	1	-	-	-	-	-	-	-	-	-	24	22	1.594	0.072
CARDINAL HEALTH - LA	LA-5119-L01	2	9	3	-	-	-	1	-	-	-	-	-	-	15	13	2.230	0.172
CARDINAL HEALTH - LA	LA-5394-L01	2	17	8	-	-	-	-	-	-	-	-	-	-	27	25	1.693	0.068
CARDINAL HEALTH - LA	LA-7096-L01	-	8	3	2	-	-	-	-	-	-	-	-	-	13	13	1.317	0.101
CARDINAL HEALTH - MA	4008	-	-	-	2	1	1	1	-	1	-	-	-	-	6	6	5.159	0.860
CARDINAL HEALTH - MA	42-0146	3	26	6	5	5	1	1	-	-	-	-	-	-	47	44	8.344	0.190
CARDINAL HEALTH - MD	MD-05-058-01	9	26	5	2	-	-	-	-	-	-	-	-	-	41	32	2.164	0.068
CARDINAL HEALTH - MD	MD-05-148-01	-	10	2	2	-	-	-	-	-	-	-	-	-	14	14	1.440	0.103
CARDINAL HEALTH - MD	MD-33-177-01	-	1	-	-	-	-	-	-	-	-	-	-	-	1	1	0.078	0.078
CARDINAL HEALTH - MD	MD-33-198-01	8	38	5	-	1	-	-	-	-	-	-	-	-	52	44	2.390	0.054
CARDINAL HEALTH - MI	1549-4	1	25	8	3	1	2	-	-	-	-	-	-	-	40	39	4.520	0.116
CARDINAL HEALTH - MN	1037-205-89	15	46	8	2	-	-	-	-	-	-	-	-	-	71	56	2.807	0.050
CARDINAL HEALTH - MN	1045-100-89	-	-	4	-	1	-	-	-	-	-	-	-	-	5	5	1.169	0.234
CARDINAL HEALTH - MO	2840	1	28	6	2	-	-	-	-	-	-	-	-	-	37	36	1.944	0.054

APPENDIX A

Table A - Annual TEDE for Agreement State Licensees

2004 (continued)

PROGRAM CODE - LICENSEE NAME	LICENSE#	No Meas. Exposure	Meas. <0.10	0.10-0.25	0.25-0.50	0.50-0.75	0.75-1.00	1.00-2.00	2.00-3.00	3.00-4.00	4.00-5.00	5.00-6.00	6.00-12.00	>12.0	Total Number Monitored	Number with Meas. Dose	Total Collective TEDE (person-rem)	Average Meas. TEDE (rem)
MANUFACTURING AND DISTRIBUTION – NUCLEAR PHARMACIES – 02500																		
CARDINAL HEALTH - NC	011-0794-7	-	8	1											9	9	0.379	0.042
CARDINAL HEALTH - NC	025-0794-10	3	5	1				1							10	7	1.544	0.221
CARDINAL HEALTH - NC	026-0794-8	3	5	2	1	1									12	9	1.405	0.156
CARDINAL HEALTH - NC	04'-0794-3	5	10												15	10	0.291	0.029
CARDINAL HEALTH - NC	041-0794-5	3	1	1											5	2	0.150	0.075
CARDINAL HEALTH - NC	06C-0794-1	-	30	5											35	35	1.751	0.050
CARDINAL HEALTH - NC	060-0794-9	1	1												2	1	0.001	0.001
CARDINAL HEALTH - NC	092-1175-1	2	8	3	2										15	13	1.293	0.099
CARDINAL HEALTH - NE	01-65-01	5	29	5	2	1									42	37	3.308	0.089
CARDINAL HEALTH - NM	RP396-02	3	10	1	1										15	12	0.650	0.054
CARDINAL HEALTH - NV	03-11-0150-01	1	11	7											19	18	1.384	0.077
CARDINAL HEALTH - NV	092-0794-6	6	15	2	2	2									27	21	2.764	0.132
CARDINAL HEALTH - NY	2306-3119MD	6	28	10		2									46	40	3.650	0.091
CARDINAL HEALTH - NY	2770-4012	1	1	1											3	2	0.112	0.056
CARDINAL HEALTH - NY	C2328	4	16	3											23	19	0.857	0.045
CARDINAL HEALTH - NY	C2364	14	19	1				1							35	21	2.268	0.108
CARDINAL HEALTH - NY	C2449	7	14	1											22	15	0.334	0.022
CARDINAL HEALTH - NY	C258E	5	15	2											22	17	0.698	0.041
CARDINAL HEALTH - NY	C2593	1	24	3											28	27	1.323	0.049
CARDINAL HEALTH - NY	C2613	2	9	1											12	10	0.518	0.052
CARDINAL HEALTH - OH	02500180000	7	22	3											32	25	0.825	0.033
CARDINAL HEALTH - OH	02500180091	3	3	1	1										8	5	0.502	0.100
CARDINAL HEALTH - OH	02500250000	5	45	7	3										60	55	2.690	0.049
CARDINAL HEALTH - OH	02500310000	4	3	2		1	1								11	7	1.893	0.270
CARDINAL HEALTH - OH	02500490001	10	29	5											44	34	1.321	0.039
CARDINAL HEALTH - OH	02500580000	3	22	3	3										31	28	1.961	0.070
CARDINAL HEALTH - OH	02500770000	1	12	1											14	13	0.409	0.031
CARDINAL HEALTH - OH	02500790000	-	9	1											10	10	0.377	0.038
CARDINAL HEALTH - OH	03214313000	-	-	1	1	1	1								4	4	1.349	0.337
CARDINAL HEALTH - OH	4853-19	6	12	8											26	20	1.383	0.069
CARDINAL HEALTH - OK	OK-19583-02MD	-	14	3	8		1								26	26	4.512	0.174
CARDINAL HEALTH - OK	OK-23359-02MD	4	17	2	1										24	20	1.062	0.053
CARDINAL HEALTH - OR	ORE-90509	4	8	4											16	12	0.732	0.061

Number of Individuals with Whole Body Doses in the Ranges (rems)

APPENDIX A

Table A - Annual TEDE for Agreement State Licensees
2004 (continued)

PROGRAM CODE - LICENSEE NAME	LICENSE#	No Meas. Exposure	Meas. <0.10	0.10-0.25	0.25-0.50	0.50-0.75	0.75-1.00	1.00-2.00	2.00-3.00	3.00-4.00	4.00-5.00	5.00-6.00	6.00-12.00	>12.0	Total Number Monitored	Number with Meas. Dose	Total Collective TEDE (person-rem)	Average Meas. TEDE (rem)
MANUFACTURING AND DISTRIBUTION – NUCLEAR PHARMACIES – 02500																		
CARDINAL HEALTH - OR	ORE-90703	3	13	4	1	-	-	-	-	-	-	-	-	-	21	18	1.395	0.078
CARDINAL HEALTH - OR	ORE-90914	4	16	6	3	1	-	-	-	-	-	-	-	-	30	26	2.876	0.111
CARDINAL HEALTH - PA	10239-02-002	2	-	-	-	-	-	-	-	-	-	-	-	-	2	-	-	-
CARDINAL HEALTH - PA	PA-0384	6	29	1	2	-	-	-	-	-	-	-	-	-	38	32	1.541	0.048
CARDINAL HEALTH - PA	PA-0385	11	51	18	11	5	1	-	-	-	-	-	-	-	97	86	12.990	0.151
CARDINAL HEALTH - PA	PA-0415	2	26	8	1	-	-	-	-	-	-	-	-	-	37	35	2.728	0.078
CARDINAL HEALTH - PA	PA-0460	1	17	1	-	-	-	-	-	-	-	-	-	-	19	18	0.456	0.025
CARDINAL HEALTH - PA	PA-0643	1	17	1	-	-	-	-	-	-	-	-	-	-	19	18	0.618	0.034
CARDINAL HEALTH - PA	PA-0680	1	17	4	2	-	-	-	-	-	-	-	-	-	24	23	2.127	0.092
CARDINAL HEALTH - PA	PA-0895	2	3	1	1	-	-	-	-	-	-	-	-	-	7	5	0.543	0.109
CARDINAL HEALTH - RI	3B-114-01	7	27	3	-	-	-	-	-	-	-	-	-	-	37	30	1.072	0.036
CARDINAL HEALTH - SC	448	4	16	3	2	1	-	-	-	-	-	-	-	-	26	22	2.689	0.122
CARDINAL HEALTH - SC	536	1	-	-	-	-	-	-	-	-	-	-	-	-	1	-	-	-
CARDINAL HEALTH - SD	47111	2	14	-	2	-	-	-	-	-	-	-	-	-	18	16	1.002	0.063
CARDINAL HEALTH - TN	R-19199-B15	5	27	9	1	-	-	-	-	-	-	-	-	-	42	37	2.682	0.072
CARDINAL HEALTH - TN	R-33111-I14	6	15	2	2	1	-	-	-	-	-	-	-	-	26	20	2.102	0.105
CARDINAL HEALTH - TN	R47080-L14	7	24	6	-	-	-	-	-	-	-	-	-	-	37	30	1.756	0.059
CARDINAL HEALTH - TN	R-57025-C15	3	5	-	-	-	-	-	-	-	-	-	-	-	8	5	0.223	0.045
CARDINAL HEALTH - TN	R-71029-B11	1	8	2	-	-	-	-	-	-	-	-	-	-	11	10	0.389	0.039
CARDINAL HEALTH - TN	R-79174-02	-	-	2	2	-	-	-	-	-	-	-	-	-	4	4	1.158	0.290
CARDINAL HEALTH - TN	R-79174-D17	1	18	2	2	1	-	-	-	-	-	-	-	-	24	23	2.133	0.093
CARDINAL HEALTH - TX	LO-1911	12	39	5	4	2	1	-	-	-	-	-	-	-	63	51	5.422	0.106
CARDINAL HEALTH - TX	LO-1999	1	9	-	-	-	-	-	-	-	-	-	-	-	10	9	0.164	0.018
CARDINAL HEALTH - TX	LO-2033	3	26	7	4	1	-	3	-	-	-	-	-	-	44	41	6.893	0.168
CARDINAL HEALTH - TX	LO-2048	23	19	6	1	-	-	-	-	-	-	-	-	-	49	26	1.748	0.067
CARDINAL HEALTH - TX	LO-2117	2	15	3	1	-	-	-	-	-	-	-	-	-	21	19	0.987	0.052
CARDINAL HEALTH - TX	LO-2737	-	3	-	-	-	-	-	-	-	-	-	-	-	4	3	0.089	0.030
CARDINAL HEALTH - TX	LO-2905	5	1	-	-	-	-	-	-	-	-	-	-	-	6	1	0.015	0.015
CARDINAL HEALTH - TX	LO-3398	1	7	-	-	-	-	-	-	-	-	-	-	-	8	7	0.185	0.026
CARDINAL HEALTH - TX	LO-4043	-	7	1	-	-	-	-	-	-	-	-	-	-	8	8	0.180	0.023
CARDINAL HEALTH - TX	LO-4573	4	25	3	1	-	-	-	-	-	-	-	-	-	33	29	2.154	0.074
CARDINAL HEALTH - TX	LO-4781	4	14	4	-	-	-	-	-	-	-	-	-	-	22	18	1.045	0.058
CARDINAL HEALTH - TX	LO-4785	3	13	4	-	-	-	-	-	-	-	-	-	-	20	17	1.079	0.063

Number of Individuals with Whole Body Doses in the Ranges (rems)*

APPENDIX A

Table A - Annual TEDE for Agreement State Licensees
2004 (continued)

PROGRAM CODE - LICENSEE NAME	LICENSE#	No Meas. Exposure	Number of Individuals with Whole Body Doses in the Ranges (rems)*												Total Number Monitored	Number with Meas. Dose	Total Collective TEDE (person-rem)	Average Meas. TEDE (rem)
			Meas. <0.10	0.10-0.25	0.25-0.50	0.50-0.75	0.75-1.00	1.00-2.00	2.00-3.00	3.00-4.00	4.00-5.00	5.00-6.00	6.00-12.00	>12.0				
MANUFACTURING AND DISTRIBUTION – NUCLEAR PHARMACIES – 02500																		
CARDINAL HEALTH - TX	L0-5461	1	3	-	-	-	-	-	-	-	-	-	-	-	4	3	0.092	0.031
CARDINAL HEALTH - TX	L0-5536	2	2	1	-	-	-	-	-	-	-	-	-	-	4	4	0.602	0.151
CARDINAL HEALTH - US	074-1180-1	-	6	1	-	-	-	-	-	-	-	-	-	-	7	7	0.462	0.066
CARDINAL HEALTH - US	CH-US-001	33	76	24	35	1	3	2	-	-	-	-	-	-	174	141	23.648	0.168
CARDINAL HEALTH - UT	UT-1800225	2	12	2	4	2	-	-	-	-	-	-	-	-	22	20	3.196	0.160
CARDINAL HEALTH - VA	760-34-1	3	44	7	-	-	-	-	-	-	-	-	-	-	54	51	1.903	0.037
CARDINAL HEALTH - WA	WN-NP003-1	5	28	3	1	-	-	-	-	-	-	-	-	-	37	32	1.419	0.044
CARDINAL HEALTH - WA	WN-NP004-1	1	22	10	6	-	-	-	-	-	-	-	-	-	39	38	4.526	0.119
CARDINAL HEALTH - WA	WN-NP005-1	4	15	1	-	-	-	-	-	-	-	-	-	-	20	16	0.690	0.043
CARDINAL HEALTH - WA	WN-NP006-1	2	4	-	-	-	-	-	-	-	-	-	-	-	6	4	0.108	0.027
CARDINAL HEALTH - WA	WN-NP011-1	6	49	9	8	3	2	-	-	-	-	-	-	-	77	71	9.950	0.140
CARDINAL HEALTH - WI	025-1123-01	1	7	3	1	-	-	1	-	-	-	-	-	-	13	12	2.654	0.221
CARDINAL HEALTH - WI	087-312-01	6	20	2	1	-	-	-	-	-	-	-	-	-	29	23	1.091	0.047
CARDINAL HEALTH - WI	141-1306-01	1	1	2	1	-	-	-	-	-	-	-	-	-	5	4	0.676	0.169
CARDINAL HEALTH - WV	0628-9	1	-	-	-	-	-	-	-	-	-	-	-	-	1	-	-	-
GE HEALTHCARE - ANAHEIM	4810-30	11	16	2	-	-	-	-	-	-	-	-	-	-	29	18	0.830	0.046
GE HEALTHCARE - SACRAMENTO	4809-34	22	7	2	-	-	-	-	-	-	-	-	-	-	31	9	0.617	0.069
GE HEALTHCARE - SAN DIEGO	5796-37	21	5	-	-	-	-	-	-	-	-	-	-	-	26	5	0.112	0.022
GE HEALTHCARE - SAN JOSE	4811-43	9	5	4	3	-	-	-	-	-	-	-	-	-	21	12	1.980	0.165
GE HEALTHCARE - VAN NUYS	5143-19	14	6	-	-	-	-	-	-	-	-	-	-	-	20	6	0.106	0.018
IBA MOLECULAR NORTH AMERICA, INC.	CA-7131-43	2	11	1	-	-	-	-	-	-	-	-	-	-	14	12	0.445	0.037
MALLINCKRODT, INC.	859-01	17	1	-	-	-	-	-	-	-	-	-	-	-	18	1	0.029	0.029
MORAVEK BIOCHEMICALS, INC.	2960-33)	-	20	2	-	-	-	-	-	-	-	-	-	-	23	23	2.611	0.114
WYLE LABORATORIES	FL-2953-1	29	3	-	-	-	-	-	-	-	-	-	-	-	32	3	0.054	0.018
Total	160	699	2,458	543	224	58	21	23	1	1	-	-	-	-	4,028	3,329	320.018	0.096
WELL LOGGING BYPRODUCT AND/OR SNM SEALED SOURCES ONLY - 03111																		
ISOTOPE PRODUCTS LABS	1509-1S	2	25	9	5	5	5	9	-	-	-	-	-	-	60	58	25.012	0.431
Total	1	2	25	9	5	5	5	9	-	-	-	-	-	-	60	58	25.012	0.431
MEASURING SYSTEMS PORTABLE GAUGES - 03121																		
CPN INTERNATIONAL, INC.	1100-07	3	3	4	3	-	-	-	-	-	-	-	-	-	14	11	2.418	0.220
Total	1	3	3	4	3	-	-	-	-	-	-	-	-	-	14	11	2.418	0.220

APPENDIX A
Table A - Annual TEDE for Agreement State Licensees
2004 (continued)

PROGRAM CODE - LICENSEE NAME	LICENSE#	No Meas. Exposure	Meas. <0.10	0.10-0.25	0.25-0.50	0.50-0.75	0.75-1.00	1.00-2.00	2.00-3.00	3.00-4.00	4.00-5.00	5.00-6.00	6.00-12.00	>12.0	Total Number Monitored	Number with Meas. Dose	Total Collective TEDE (person-rem)	Average Meas. TEDE (rem)
MANUFACTURING AND DISTRIBUTION – OTHER - 03214																		
HOCHIKI AMERICA CORP	2090-30	17	-	1	-	-	-	-	-	-	-	-	-	-	18	1	0.179	0.179
J. L. SHEPHERD AND ASSOCIATES	CA 1777-19	22	2	1	2	-	-	-	-	-	-	-	-	-	27	5	0.952	0.190
Total	2	39	2	2	2	-	-	-	-	-	-	-	-	-	45	6	1.131	0.189
INDUSTRIAL RADIOGRAPHY – FIXED LOCATION – 03310																		
LAFAYETTE TESTING SERVICES, INC.	079-1147-01	1	3	-	-	-	-	-	-	-	-	-	-	-	4	3	0.162	0.054
METALTEK - WISCONSIN CENTRIFUGAL DIVISION	133-1181-01	7	1	1	3	-	-	-	-	-	-	-	-	-	12	5	1.305	0.261
SONGS - RADIOGRAPHY GROUP	CA-5244-30	12	-	-	-	-	-	-	-	-	-	-	-	-	12	-	-	-
WAUKESHA FOUNDRY, INC. - WI	133-1337-01	3	-	-	-	-	-	-	-	-	-	-	-	-	3	-	-	-
Total	4	23	4	1	3	-	-	-	-	-	-	-	-	-	31	8	1.467	0.183
INDUSTRIAL RADIOGRAPHY – TEMPORARY JOB SITE – 03320																		
ACUREN	133-2008-01	3	9	7	8	7	6	12	7	-	-	-	-	-	59	56	45.674	0.816
ALPHA-OMEGA SERVICES, INC.	3925-19	2	7	4	2	1	-	-	-	-	-	-	-	-	16	14	1.987	0.142
AMERICAN ENGINEERING TESTING, INC.	073-2012-02	-	7	3	3	-	8	-	2	2	-	-	-	-	27	27	26.472	0.980
ANVIL INTERNATIONAL, INC.	RI-3D-064-01	1	3	-	1	-	-	-	-	-	-	-	-	-	5	4	0.241	0.060
CONSTRUCTION MATERIALS TESTING, INC.	0799-07	4	-	1	1	-	-	-	-	-	-	-	-	-	6	2	0.544	0.272
COOPER-HEAT-MQS INSPECTION	388-01	-	4	1	4	-	2	4	-	-	-	-	-	-	15	15	7.974	0.532
EDGE INSPECTION GROUP, INC.	CA-7214-48	-	3	1	-	-	1	1	-	-	-	-	-	-	6	6	2.860	0.477
FTS, INC.	OK-31027-01	-	3	6	5	3	1	3	-	-	-	-	-	-	21	21	11.675	0.556
GREAT LAKES TESTING, INC	009-1116-01	3	3	2	1	1	1	-	1	-	-	-	-	-	12	9	3.199	0.355
INTERNATIONAL INSPECTION, INC.	5046-19	1	3	2	2	1	1	-	-	-	-	-	-	-	10	9	2.894	0.322
NATIONAL SERVICE DEVELOPMENT GROUP	6313-30	-	4	1	1	1	1	1	-	-	-	-	-	-	9	9	3.216	0.357
NDE SERVICES, INC.	406-01	1	4	1	2	1	-	1	-	-	-	-	-	-	10	9	2.618	0.291
NDT LABORATORIES, INC.	1651-43	-	1	1	-	1	-	-	-	-	-	-	-	-	3	3	0.494	0.165
NDT SPECIALISTS, INC.	079-1199-01	1	4	1	2	-	1	-	-	-	-	-	-	-	13	12	5.611	0.468
TC INSPECTION, INC.	5299-07	17	17	5	6	8	2	8	1	-	-	-	-	-	64	47	23.631	0.503
TEAM INDUSTRIAL SERVICES, INC.	079-2005-01	2	5	2	1	1	-	2	1	-	-	-	-	-	14	12	4.709	0.392
TESTING ENGINEERS, INC.	1898-01	-	2	1	-	-	2	-	-	-	-	-	-	-	5	5	3.922	0.784
TWIN PORTS TESTING, INC.	031-1317-02	4	5	1	2	3	-	-	-	-	-	-	-	-	15	11	3.690	0.335
WESTERN INDUSTRIAL X-RAY	4424-48	-	1	-	3	1	-	1	-	-	-	-	-	-	6	6	4.408	0.735
WESTERN X-RAY CORP	5324-56	1	2	-	1	1	-	-	-	-	-	-	-	-	5	4	2.628	0.657
Total	20	40	87	43	44	32	16	45	12	2	-	-	-	-	321	281	158.447	0.564
IRRADIATORS - OTHER - LESS THAN 10000 CURIES - 03511																		
INDUSTRIAL NUCLEAR CO., INC.	2229-01	8	1	-	1	-	1	1	-	-	-	-	-	-	12	4	2.291	0.573
Total	1	8	1	-	1	-	1	1	-	-	-	-	-	-	12	4	2.291	0.573

APPENDIX A

Table A - Annual TEDE for Agreement State Licensees
2005

PROGRAM CODE - LICENSEE NAME	LICENSE#	No Meas. Exposure	Meas. <0.10	0.10-0.25	0.25-0.50	0.50-0.75	0.75-1.00	1.00-2.00	2.00-3.00	3.00-4.00	4.00-5.00	5.00-6.00	6.00-12.00	>12.0	Total Number Monitored	Number with Meas. Dose	Total Collective TEDE (person-rem)	Average Meas. TEDE (rem)
MANUFACTURING AND DISTRIBUTION – NUCLEAR PHARMACIES – 02500																		
CARDINAL HEALTH - AK	0707-3A907-912	3	4	1	-	-	-	-	-	-	-	-	-	-	8	5	0.280	0.056
CARDINAL HEALTH - AL	1068	3	7	1	-	-	-	-	-	-	-	-	-	-	11	8	0.255	0.032
CARDINAL HEALTH - AL	1168	1	42	11	2	-	-	-	-	-	-	-	-	-	56	55	4.437	0.081
CARDINAL HEALTH - AR	ARK-642-02500	11	60	6	2	-	-	-	-	-	-	-	-	-	79	68	3.889	0.057
CARDINAL HEALTH - AZ	01-084	3	8	1	1	-	-	-	-	-	-	-	-	-	13	10	0.949	0.095
CARDINAL HEALTH - AZ	07-123	24	30	12	3	-	-	-	-	-	-	-	-	-	69	45	3.244	0.072
CARDINAL HEALTH - CA	2891-57	1	12	2	3	-	-	-	-	-	-	-	-	-	18	17	1.673	0.098
CARDINAL HEALTH - CA	3317-19	1	15	2	1	-	-	-	-	-	-	-	-	-	19	18	1.932	0.107
CARDINAL HEALTH - CA	3426-43	-	8	1	-	-	-	-	-	-	-	-	-	-	9	9	0.401	0.045
CARDINAL HEALTH - CA	3469-1	-	-	-	1	1	-	-	-	-	-	-	-	-	2	2	0.938	0.469
CARDINAL HEALTH - CA	3469-2	1	3	-	-	1	-	-	3	1	-	-	-	-	9	8	10.859	1.357
CARDINAL HEALTH - CA	3673-34	13	31	4	-	-	-	-	-	-	-	-	-	-	48	35	1.143	0.033
CARDINAL HEALTH - CA	3822-19	3	33	17	4	1	-	-	-	-	-	-	-	-	58	55	5.929	0.108
CARDINAL HEALTH - CA	3832-01	1	21	6	3	-	-	1	-	-	-	-	-	-	32	31	4.215	0.136
CARDINAL HEALTH - CA	4905-10	7	26	-	1	-	-	-	-	-	-	-	-	-	34	27	1.023	0.038
CARDINAL HEALTH - CA	4999-30	6	25	7	-	-	-	-	-	-	-	-	-	-	38	32	1.880	0.059
CARDINAL HEALTH - CA	5218-36	-	10	2	4	2	-	1	-	-	-	-	-	-	19	19	4.596	0.242
CARDINAL HEALTH - CA	5905-15	-	6	-	-	-	-	-	-	-	-	-	-	-	6	6	0.136	0.023
CARDINAL HEALTH - CA	5910-50	-	5	2	-	-	-	-	-	-	-	-	-	-	7	7	0.355	0.051
CARDINAL HEALTH - CA	6321-45	4	3	-	-	-	-	-	-	-	-	-	-	-	7	3	0.119	0.040
CARDINAL HEALTH - CA	6691-33	-	10	-	-	-	-	-	-	-	-	-	-	-	10	10	0.127	0.013
CARDINAL HEALTH - CA	6924-36	-	-	-	-	1	2	1	-	-	-	-	-	-	4	4	3.709	0.927
CARDINAL HEALTH - CA	6925-19	-	4	2	3	-	1	-	-	-	-	-	-	-	10	10	2.474	0.247
CARDINAL HEALTH - CO	162-06	3	9	2	3	1	3	1	-	-	-	-	-	-	21	18	4.624	0.257
CARDINAL HEALTH - CO	392-01	39	34	9	6	1	-	-	-	-	-	-	-	-	89	50	4.709	0.094
CARDINAL HEALTH - CO	392-03	11	27	6	5	1	-	-	-	-	-	-	-	-	50	39	3.507	0.090
CARDINAL HEALTH - CT	00163	4	26	9	3	-	-	-	-	-	-	-	-	-	42	38	3.533	0.093
CARDINAL HEALTH - FL	12264-9	3	15	3	-	-	-	-	-	-	-	-	-	-	21	18	0.619	0.034
CARDINAL HEALTH - FL	3273-7	7	12	1	1	-	-	-	-	-	-	-	-	-	21	14	1.197	0.086
CARDINAL HEALTH - FL	3453-1	-	14	2	1	-	-	-	-	-	-	-	-	-	17	17	0.919	0.054
CARDINAL HEALTH - FL	3453-2	12	13	4	-	-	-	-	-	-	-	-	-	-	29	17	0.867	0.051
CARDINAL HEALTH - FL	3453-3	4	17	3	-	-	-	-	-	-	-	-	-	-	24	20	0.741	0.037
CARDINAL HEALTH - FL	3453-5	6	11	3	3	-	2	-	-	-	-	-	-	-	25	19	3.402	0.179

APPENDIX A

Table A - Annual TEDE for Agreement State Licensees

2005 (continued)

PROGRAM CODE - LICENSEE NAME	LICENSE#	No Meas. Exposure	Meas. <0.10	Number of Individuals with Whole Body Doses in the Ranges (rems)*												Total Number Monitored	Number with Meas. Dose	Total Collective TEDE (person-rem)	Average Meas. TEDE (rem)
				0.10-0.25	0.25-0.50	0.50-0.75	0.75-1.00	1.00-2.00	2.00-3.00	3.00-4.00	4.00-5.00	5.00-6.00	6.00-12.00	>12.0				
MANUFACTURING AND DISTRIBUTION – NUCLEAR PHARMACIES – 02500																		
CARDINAL HEALTH - FL	3453-6	4	19	5	3	1	-	-	-	-	-	-	-	-	32	28	3.223	0.115
CARDINAL HEALTH - FL	3453-7	4	15	8	4	2	-	-	-	-	-	-	-	-	33	29	4.673	0.161
CARDINAL HEALTH - FL	3453-8	6	22	4	4	-	-	-	-	-	-	-	-	-	36	30	2.582	0.086
CARDINAL HEALTH - FL	3453-9	3	17	3	-	-	-	-	-	-	-	-	-	-	23	20	0.855	0.043
CARDINAL HEALTH - FL	FL-3273-3	1	5	-	-	-	-	-	-	-	-	-	-	-	6	5	0.023	0.005
CARDINAL HEALTH - FL	FL-3273-6	1	-	-	-	-	-	-	-	-	-	-	-	-	1	-	-	-
CARDINAL HEALTH - GA	GA-467-1MD	4	20	5	3	1	-	-	-	-	-	-	-	-	33	29	3.076	0.106
CARDINAL HEALTH - GA	GA-467-2MD	3	10	-	-	-	-	-	-	-	-	-	-	-	13	10	0.188	0.019
CARDINAL HEALTH - GA	GA-467-3MD	3	2	-	-	-	-	-	-	-	-	-	-	-	5	2	0.030	0.015
CARDINAL HEALTH - GA	GA-823-2MD	1	13	2	1	-	-	-	-	-	-	-	-	-	17	16	1.346	0.084
CARDINAL HEALTH - IA	0043-1-77-NP	5	14	-	-	-	-	-	-	-	-	-	-	-	19	14	0.375	0.027
CARDINAL HEALTH - IL	IL-01721-01	5	58	7	-	-	-	-	-	-	-	-	-	-	70	65	2.622	0.040
CARDINAL HEALTH - IN	R0358-45	9	24	2	-	-	-	-	-	-	-	-	-	-	35	26	0.872	0.034
CARDINAL HEALTH - KS	20-C495-01	6	14	4	2	-	-	-	-	-	-	-	-	-	26	20	1.564	0.078
CARDINAL HEALTH - KY	202-204-32	12	28	3	-	-	-	-	-	-	-	-	-	-	43	31	1.232	0.040
CARDINAL HEALTH - KY	202-206-32	-	23	5	2	-	-	-	-	-	-	-	-	-	30	30	2.276	0.076
CARDINAL HEALTH - LA	LA-10217-L01	4	3	1	-	-	-	-	-	-	-	-	-	-	8	4	0.212	0.053
CARDINAL HEALTH - LA	LA-10336-L01	2	9	1	-	-	-	-	-	-	-	-	-	-	12	10	0.424	0.042
CARDINAL HEALTH - LA	LA-3385-L01	23	16	3	-	-	-	-	-	-	-	-	-	-	42	19	0.844	0.044
CARDINAL HEALTH - LA	LA-5115-L01	3	21	4	-	-	-	-	-	-	-	-	-	-	28	25	1.364	0.055
CARDINAL HEALTH - LA	LA-5119-L01	4	10	-	-	-	-	-	-	-	-	-	-	-	14	10	0.245	0.025
CARDINAL HEALTH - LA	LA-5394-L01	2	18	5	2	1	-	-	-	-	-	-	-	-	28	26	3.012	0.116
CARDINAL HEALTH - LA	LA-7096-L01	1	8	2	2	-	-	-	-	-	-	-	-	-	13	12	1.028	0.086
CARDINAL HEALTH - MA	4008	-	1	-	1	1	-	1	-	1	-	-	-	-	5	5	4.940	0.988
CARDINAL HEALTH - MA	42-0146	3	24	15	5	5	2	1	-	-	-	-	-	-	55	52	11.261	0.217
CARDINAL HEALTH - MD	MD-05-058-01	7	22	-	-	-	-	-	-	-	-	-	-	-	29	22	0.531	0.024
CARDINAL HEALTH - MD	MD-05-148-01	2	5	5	2	1	-	-	-	-	-	-	-	-	15	13	2.313	0.178
CARDINAL HEALTH - MD	MD-33-177-01	-	1	-	-	1	1	-	-	-	-	-	-	-	3	3	1.901	0.634
CARDINAL HEALTH - MD	MD-33-198-01	5	31	12	5	3	1	-	-	-	-	-	-	-	57	52	7.288	0.140
CARDINAL HEALTH - MI	1549-4	7	19	5	1	3	1	-	-	-	-	-	-	-	36	29	4.642	0.160
CARDINAL HEALTH - MN	1037-205-89	16	38	10	3	1	1	-	-	-	-	-	-	-	68	52	4.241	0.082
CARDINAL HEALTH - MN	1045-100-89	-	-	1	1	-	1	-	-	-	-	-	-	-	3	3	1.249	0.416
CARDINAL HEALTH - MO	2840	4	29	7	1	-	-	-	-	-	-	-	-	-	41	37	2.300	0.062

APPENDIX A
Table A - Annual TEDE for Agreement State Licensees
2005 (continued)

MANUFACTURING AND DISTRIBUTION – NUCLEAR PHARMACIES – 02500

PROGRAM CODE - LICENSEE NAME	LICENSE#	No Meas. Exposure	Meas. <0.10	0.10-0.25	0.25-0.50	0.50-0.75	0.75-1.00	1.00-2.00	2.00-3.00	3.00-4.00	4.00-5.00	5.00-6.00	6.00-12.00	>12.0	Total Number Monitored	Number with Meas. Dose	Total Collective TEDE (person-rem)	Average Meas. TEDE (rem)
CARDINAL HEALTH - MO	6141		1				2	1	-	-	-	-	-	-	4	4	3.079	0.770
CARDINAL HEALTH - MS	MS-493-01	2	14	2	2	1			-	-	-	-	-	-	21	19	2.182	0.115
CARDINAL HEALTH - MS	MS-731-01		5	2	2		1		-	-	-	-	-	-	10	10	2.178	0.218
CARDINAL HEALTH - MS	MS-924-01	1	6	1	4				-	-	-	-	-	-	12	11	1.805	0.164
CARDINAL HEALTH - NC	011-0794-7	2	8		3				-	-	-	-	-	-	13	11	1.369	0.124
CARDINAL HEALTH - NC	025-0794-10	2	5	5	2	1		1	-	-	-	-	-	-	16	14	3.847	0.275
CARDINAL HEALTH - NC	026-0794-8	2	7	3	3	2		1	-	-	-	-	-	-	18	16	6.178	0.386
CARDINAL HEALTH - NC	041-0794-3	1	10	2	1				-	-	-	-	-	-	14	13	0.814	0.063
CARDINAL HEALTH - NC	041-0734-5	6	7		1				-	-	-	-	-	-	14	8	0.408	0.051
CARDINAL HEALTH - NC	060-0734-1	6	27	2	3				-	-	-	-	-	-	38	32	1.869	0.058
CARDINAL HEALTH - NC	060-0794-9	2							-	-	-	-	-	-	2			
CARDINAL HEALTH - NC	092-1175-1	2	4						-	-	-	-	-	-	6	4	0.107	0.027
CARDINAL HEALTH - NE	01-65-01	5	29	5	2				-	-	-	-	-	-	41	36	2.224	0.062
CARDINAL HEALTH - NM	RP-396-02	5	7		1				-	-	-	-	-	-	13	8	0.397	0.050
CARDINAL HEALTH - NV	03-11-0:50-01	4	14	5					-	-	-	-	-	-	23	19	1.247	0.066
CARDINAL HEALTH - NV	092-0794-6	6	23	6	3	1		1	-	-	-	-	-	-	40	34	4.260	0.125
CARDINAL HEALTH - NY	2306-3119MD	19	17	2	1				-	-	-	-	-	-	39	20	2.004	0.100
CARDINAL HEALTH - NY	C2328	13	6	3					-	-	-	-	-	-	22	9	0.619	0.069
CARDINAL HEALTH - NY	C2364	13	22	3	1				-	-	-	-	-	-	39	26	1.120	0.043
CARDINAL HEALTH - NY	C2449	12	13	1					-	-	-	-	-	-	26	14	0.328	0.023
CARDINAL HEALTH - NY	C2588	4	16	3					-	-	-	-	-	-	23	19	0.900	0.047
CARDINAL HEALTH - NY	C2593	9	22		1				-	-	-	-	-	-	32	23	0.905	0.039
CARDINAL HEALTH - NY	C2613	5	17	3					-	-	-	-	-	-	25	20	0.898	0.045
CARDINAL HEALTH - NY	C3046	1	2						-	-	-	-	-	-	3	2	0.028	0.014
CARDINAL HEALTH - NY	C3153	1	3						-	-	-	-	-	-	4	3	0.193	0.064
CARDINAL HEALTH - OH	02500180000	7	19	2	1				-	-	-	-	-	-	29	22	1.164	0.053
CARDINAL HEALTH - OH	02500180091		2	1		1			-	-	-	-	-	-	4	4	1.086	0.272
CARDINAL HEALTH - OH	02500250000	6	43	12	2				-	-	-	-	-	-	63	57	3.496	0.061
CARDINAL HEALTH - OH	02500310090	2	8	3		1			-	-	-	-	-	-	14	12	1.628	0.136
CARDINAL HEALTH - OH	02500490OC1	14	23	3	1				-	-	-	-	-	-	41	27	1.239	0.046
CARDINAL HEALTH - OH	02500580000	3	24	7					-	-	-	-	-	-	34	31	1.541	0.050
CARDINAL HEALTH - OH	02500770000		8	1					-	-	-	-	-	-	9	9	0.239	0.027
CARDINAL HEALTH - OH	02500790000	2	9						-	-	-	-	-	-	11	9	0.263	0.029

Number of Individuals with Whole Body Doses in the Ranges (rems)

APPENDIX A
Table A - Annual TEDE for Agreement State Licensees 2005 (continued)

PROGRAM CODE - LICENSEE NAME	LICENSE#	No Meas. Exposure	Meas. <0.10	0.10-0.26	0.25-0.50	0.50-0.75	0.75-1.00	1.00-2.00	2.00-3.00	3.00-4.00	4.00-5.00	5.00-6.00	6.00-12.00	>12.0	Total Number Monitored	Number with Meas. Dose	Total Collective TEDE (person-rem)	Average Meas. TEDE (rem)
MANUFACTURING AND DISTRIBUTION – NUCLEAR PHARMACIES – 02500																		
CARDINAL HEALTH - OH	03214310000	-	-	1	3	-	-	-	-	-	-	-	-	-	4	4	1.481	0.370
CARDINAL HEALTH - OH	4853-19	5	17	6	1	-	-	-	-	-	-	-	-	-	29	24	1.854	0.077
CARDINAL HEALTH - OK	OK-19583-02MD	1	20	3	3	-	-	-	-	-	-	-	-	-	27	26	1.946	0.075
CARDINAL HEALTH - OK	OK-23359-02MD	7	13	1	-	-	-	-	-	-	-	-	-	-	21	14	0.400	0.029
CARDINAL HEALTH - OR	ORE-90509	3	8	4	-	-	-	-	-	-	-	-	-	-	15	12	0.703	0.059
CARDINAL HEALTH - OR	ORE-90703	6	16	3	2	-	-	-	-	-	-	-	-	-	27	21	1.640	0.078
CARDINAL HEALTH - OR	ORE-90914	11	22	9	3	-	-	-	-	-	-	-	-	-	45	34	3.446	0.101
CARDINAL HEALTH - PA	10239-02-002	12	14	3	1	-	-	-	-	-	-	-	-	-	30	18	1.309	0.073
CARDINAL HEALTH - PA	PA-0384	12	31	3	3	-	-	-	-	-	-	-	-	-	49	37	2.120	0.057
CARDINAL HEALTH - PA	PA-0385	17	43	9	9	4	-	-	-	-	-	-	-	-	82	65	8.744	0.135
CARDINAL HEALTH - PA	PA-0415	3	22	10	-	-	-	-	-	-	-	-	-	-	35	32	3.011	0.094
CARDINAL HEALTH - PA	PA-0460	10	11	-	1	-	-	-	-	-	-	-	-	-	22	12	0.561	0.047
CARDINAL HEALTH - PA	PA-0643	6	10	-	-	-	-	-	-	-	-	-	-	-	16	10	0.353	0.035
CARDINAL HEALTH - PA	PA-0680	1	21	1	2	-	1	-	-	-	-	-	-	-	26	25	2.149	0.086
CARDINAL HEALTH - PA	PA-0895	1	2	-	1	-	-	-	-	-	-	-	-	-	4	3	0.352	0.117
CARDINAL HEALTH - RI	3B-114-01	7	31	-	-	-	-	-	-	-	-	-	-	-	38	31	0.860	0.028
CARDINAL HEALTH - SC	448	1	14	7	2	-	-	-	-	-	-	-	-	-	24	23	2.517	0.109
CARDINAL HEALTH - SD	47111	3	14	2	1	-	-	-	-	-	-	-	-	-	20	17	1.172	0.069
CARDINAL HEALTH - TN	R-19199-B15	10	27	7	-	-	-	-	-	-	-	-	-	-	44	34	1.729	0.051
CARDINAL HEALTH - TN	R-33111-I14	1	21	2	1	-	-	-	-	-	-	-	-	-	25	24	0.938	0.039
CARDINAL HEALTH - TN	R47080-L14	12	21	3	1	-	-	-	-	-	-	-	-	-	37	25	1.564	0.063
CARDINAL HEALTH - TN	R-57025-C15	5	5	-	-	-	-	-	-	-	-	-	-	-	10	5	0.114	0.023
CARDINAL HEALTH - TN	R-71029-B11	-	7	-	-	-	-	-	-	-	-	-	-	-	7	7	0.240	0.034
CARDINAL HEALTH - TN	R-79174-D17	2	12	1	7	1	-	-	-	-	-	-	-	-	23	21	3.642	0.173
CARDINAL HEALTH - TX	L0-1911	16	45	3	9	-	-	-	-	-	-	-	-	-	73	57	4.800	0.084
CARDINAL HEALTH - TX	L0-1999	2	6	-	-	-	-	-	-	-	-	-	-	-	8	6	0.089	0.015
CARDINAL HEALTH - TX	L0-2033	12	25	7	3	3	-	-	-	-	-	-	-	-	51	39	5.431	0.139
CARDINAL HEALTH - TX	L0-2048	22	13	4	1	-	-	-	-	-	-	-	-	-	40	18	1.515	0.084
CARDINAL HEALTH - TX	L0-2117	3	17	3	-	-	-	-	-	-	-	-	-	-	23	20	0.849	0.042
CARDINAL HEALTH - TX	L0-2737	2	1	1	-	-	-	-	-	-	-	-	-	-	4	2	0.126	0.063
CARDINAL HEALTH - TX	L0-3398	2	6	-	-	-	-	-	-	-	-	-	-	-	8	6	0.156	0.026
CARDINAL HEALTH - TX	L0-4043	3	6	1	-	-	-	-	-	-	-	-	-	-	10	7	0.179	0.026
CARDINAL HEALTH - TX	L0-4573	7	24	11	3	-	-	-	-	-	-	-	-	-	45	38	3.301	0.087

*Number of Individuals with Whole Body Doses in the Ranges (rems)

APPENDIX A
Table A - Annual TEDE for Agreement State Licensees
2005 (continued)

PROGRAM CODE - LICENSEE NAME	LICENSE#	No Meas. Exposure	Meas. <0.10	0.10-0.25	0.25-0.50	0.50-0.75	0.75-1.00	1.00-2.00	2.00-3.00	3.00-4.00	4.00-5.00	5.00-6.00	6.00-12.00	>12.0	Total Number Monitored	Number with Meas. Dose	Total Collective TEDE (person-rem)	Average Meas. TEDE (rem)
MANUFACTURING AND DISTRIBUTION – NUCLEAR PHARMACIES – 02500																		
CARDINAL HEALTH - TX	L0-4781	9	11	2	-	-	-	-	-	-	-	-	-	-	22	13	0.503	0.039
CARDINAL HEALTH - TX	L0-4785	2	13	3	-	-	-	-	-	-	-	-	-	-	18	16	0.576	0.036
CARDINAL HEALTH - TX	L0-5461	3	5	-	-	-	-	-	-	-	-	-	-	-	8	5	0.124	0.025
CARDINAL HEALTH - TX	L0-5536	1	1	1	-	-	-	2	-	-	-	-	-	-	5	4	1.907	0.477
CARDINAL HEALTH - US	065-0794.11	-	2	-	-	-	-	-	-	-	-	-	-	-	2	2	0.107	0.054
CARDINAL HEALTH - US	CH-US-0C1	24	56	27	20	5	1	2	-	-	-	-	-	-	135	111	20.303	0.183
CARDINAL HEALTH - UT	UT-1800225	2	19	4	-	-	-	-	-	-	-	-	-	-	25	23	1.700	0.074
CARDINAL HEALTH - VA	760-34-1	10	47	1	4	-	-	-	-	-	-	-	-	-	62	52	2.373	0.046
CARDINAL HEALTH - WA	WN-NP003-1	6	26	5	2	-	-	-	-	-	-	-	-	-	39	33	2.511	0.076
CARDINAL HEALTH - WA	WN-NP004-1	1	11	11	4	1	-	-	-	-	-	-	-	-	28	27	3.991	0.148
CARDINAL HEALTH - WA	WN-NP005-1	1	24	1	-	-	-	-	-	-	-	-	-	-	26	25	1.104	0.044
CARDINAL HEALTH - WA	WN-NP006-1	1	5	1	-	-	-	-	-	-	-	-	-	-	7	6	0.342	0.057
CARDINAL HEALTH - WA	WN-NP011-1	15	41	7	10	5	3	-	-	-	-	-	-	-	81	66	11.721	0.178
CARDINAL HEALTH - WI	025-1123-C1	4	10	1	-	-	1	-	-	-	-	-	-	-	16	12	2.284	0.190
CARDINAL HEALTH - WI	087-1312-C1	6	16	3	-	-	-	-	-	-	-	-	-	-	25	19	0.828	0.044
CARDINAL HEALTH - WI	141-1306-C1	3	8	-	-	-	-	-	-	-	-	-	-	-	11	8	0.238	0.030
GE HEALTHCARE - ANAHEIM	4810-30	14	9	5	-	-	-	-	-	-	-	-	-	-	28	14	0.922	0.066
GE HEALTHCARE - SACRAMENTO	4809-34	29	6	1	-	-	-	-	-	-	-	-	-	-	36	7	0.527	0.075
GE HEALTHCARE - SAN DIEGO	5796-37	24	3	-	-	-	-	-	-	-	-	-	-	-	27	3	0.071	0.024
GE HEALTHCARE - SAN JOSE	4811-43	7	5	8	3	-	-	-	-	-	-	-	-	-	23	16	2.424	0.152
GE HEALTHCARE - VAN NUYS	5143-19	13	7	-	-	-	-	-	-	-	-	-	-	-	20	7	0.155	0.022
IBA MOLECULAR NORTH AMERICA, INC.	CA-7131-43	4	11	1	2	2	1	-	-	-	-	-	-	-	21	17	3.355	0.197
MALLINCKRODT, INC.	859-01	20	4	-	-	-	-	-	-	-	-	-	-	-	24	4	0.102	0.026
MORAVEK BIOCHEMICALS, INC	2960-30	15	8	6	2	-	-	-	-	-	-	-	-	-	23	23	2.255	0.098
Total	**156**	882	2,401	506	229	58	22	17	4	2	-	-	-	-	4,121	3,239	320.742	0.099
WELL LOGGING BYPRODUCT AND/OR SNM SEALED SOURCES ONLY - 03111																		
ISOTOPE PRODUCTS LABS	1509-19	7	23	8	7	5	6	8	1	-	-	-	-	-	65	58	25.165	0.434
Total	**1**	7	23	8	7	5	6	8	1	-	-	-	-	-	65	58	25.165	0.434
MEASURING SYSTEMS PORTABLE GAUGES - C3121																		
CPN INTERNATIONAL, INC.	1100-07	1	7	4	1	1	-	-	-	-	-	-	-	-	14	13	1.656	0.127
Total	**1**	7	4	1	1	-	-	-	-	-	-	-	-	-	14	13	1.656	0.127

APPENDIX A
Table A - Annual TEDE for Agreement State Licensees
2005 (continued)

PROGRAM CODE - LICENSEE NAME	LICENSE#	No Meas. Exposure	Number of Individuals with Whole Body Doses in the Ranges (rems)*												Total Number Monitored	Number with Meas. Dose	Total Collective TEDE (person-rem)	Average Meas. TEDE (rem)
			Meas. <0.10	0.10-0.25	0.25-0.50	0.50-0.75	0.75-1.00	1.00-2.00	2.00-3.00	3.00-4.00	4.00-5.00	5.00-6.00	6.00-12.00	>12.0				
MANUFACTURING AND DISTRIBUTION - OTHER - 03214																		
HOCHIKI AMERICA CORP	2090-30	19	-	-	-	-	-	-	-	-	-	-	-	-	19		-	-
J.L. SHEPHERD AND ASSOCIATES	CA 1777-19	16	4	2	2	1	-	-	-	-	-	-	-	-	25	9	1.734	0.193
Total	2	35	4	2	2	1	-	-	-	-	-	-	-	-	44	9	1.734	0.193
WASTE DISPOSAL SERVICE PROCESSING AND/OR REPACKAGING - 03234																		
ENVIRONMENTAL MANAGEMENT & CONTROLS	3546-50	-	1	1	-	-	1	2	-	-	-	-	-	-	5	5	3.120	0.624
Total	1	-	1	1	-	-	1	2	-	-	-	-	-	-	5	5	3.120	0.624
INDUSTRIAL RADIOGRAPHY - FIXED LOCATION - 03310																		
LAFAYETTE TESTING SERVICES, INC.	079-1147-01	-	7	1	-	-	-	-	-	-	-	-	-	-	8	8	0.596	0.075
METALTEK - WISCONSIN CENTRIFUGAL DIVISION	133-1181-01	3	2	2	1	-	-	-	-	-	-	-	-	-	8	5	0.934	0.187
SONGS - RADIOGRAPHY GROUP	CA-5244-30	8	-	-	-	-	-	-	-	-	-	-	-	-	8	-	-	-
WAUKESHA FOUNDRY, INC. - WI	133-1337-01	3	-	-	-	-	-	-	-	-	-	-	-	-	3	-	-	-
Total	4	14	9	3	1	-	-	-	-	-	-	-	-	-	27	13	1.530	0.118
INDUSTRIAL RADIOGRAPHY - TEMPORARY JOB SITE - 03320																		
ACUREN	133-2008-01	2	4	10	2	1	3	10	1	-	-	-	-	-	33	31	23.755	0.766
AITEC USA, INC.	L05718	5	14	15	9	2	4	3	-	-	-	-	-	-	52	47	14.440	0.307
ALPHA-OMEGA SERVICES, INC.	3925-19	2	7	3	1	1	1	1	-	-	-	-	-	-	16	14	7.093	0.507
AMERICAN ENGINEERING TESTING, INC.	073-2012-02	-	2	2	2	1	1	4	-	1	-	-	-	-	13	13	16.539	1.272
ANVIL INTERNATIONAL, INC.	RI-3D-064-01	1	3	1	-	-	-	-	-	-	-	-	-	-	5	4	0.271	0.068
CONSTRUCTION MATERIALS TESTING, INC.	0799-07	1	4	-	1	-	1	-	-	-	-	-	-	-	7	6	1.052	0.175
COOPERHEAT-MQS INSPECTION	388-01	1	3	2	-	2	1	1	-	-	-	-	-	-	10	9	3.467	0.385
EDGE INSPECTION GROUP, INC.	CA-7214-48	-	3	-	1	-	2	-	-	-	-	-	-	-	6	6	2.938	0.490

APPENDIX A

Table A - Annual TEDE for Agreement State Licensees

2005 (continued)

PROGRAM CODE - LICENSEE NAME	LICENSE#	No Meas. Exposure	Number of Individuals with Whole Body Doses in the Ranges (rems)*												Total Number Monitored	Number with Meas. Dose	Total Collective TEDE (person-rem)	Average Meas. TEDE (rem)
			Meas. <0.10	0.10-0.25	0.25-0.50	0.50-0.75	0.75-1.00	1.00-2.00	2.00-3.00	3.00-4.00	4.00-5.00	5.00-6.00	6.00-12.00	>12.0				
GREAT LAKES TESTING, INC.	009-1116-01	9	7	1	3	2	-	-	-	-	-	-	-	-	22	13	2.667	0.205
INTERNATIONAL INSPECTION, INC.	5046-19	2	2	3	1	2	-	-	-	-	-	-	-	-	10	8	2.048	0.256
IRISNDT, INC.	L-04769	3	16	6	17	7	10	13	16	4	3	1	-	-	96	93	114.077	1.227
NATIONAL SERVICE DEVELOPMENT GROUP	6313-3C	1	5	2	2	1	1	2	-	-	-	-	-	-	14	13	5.338	0.411
NDE SERVICES, INC	406-01	-	4	1	1	2	-	-	-	-	-	-	-	-	8	8	1.837	0.230
NDTLABORATORIES, INC.	1651-43	3	1	-	2	-	-	-	-	-	-	-	-	-	6	3	0.938	0.313
NDT SPECIALISTS, INC.	079-1199-01	-	6	1	4	-	-	2	-	-	-	-	-	-	13	13	6.750	0.519
TC INSPECTION, INC.	5299-07	22	16	12	5	6	8	-	2	-	-	-	-	-	71	49	23.177	0.473
TEAM INDUSTRIAL SERVICES, INC.	079-2005-01	1	4	3	2	2	1	-	-	-	-	-	-	-	13	12	3.708	0.309
TESTING ENGINEERS, INC.	1898-01	-	2	2	2	-	1	-	-	-	-	-	-	-	7	7	2.205	0.315
TWIN PORTS TESTING, INC.	031-1317-02	6	3	1	2	2	1	-	-	-	-	-	-	-	15	9	4.130	0.459
WESTERN INDUSTRIAL X-RAY	4424-48	-	-	5	3	2	2	2	-	-	-	-	-	-	14	14	8.019	0.573
WESTERN X-RAY CORP	5324-56	1	3	1	-	1	-	-	-	-	-	-	-	-	6	5	0.992	0.198
YUBA HEAT TRANSFER	OK-1-3735-01	-	1	1	-	-	-	-	-	-	-	-	-	-	2	2	0.163	0.082
Total	22	60	110	72	59	34	25	44	25	5	4	1	-	-	439	379	245.604	0.648
IRRADIATORS - OTHER - LESS THAN 10000 CURIES - 03511																		
INDUSTRIAL NUCLEAR CO., INC.	2229-01	5	1	-	1	1	-	1	-	-	-	-	-	-	9	4	2.232	0.558
Total	1	5	1	-	1	1	-	1	-	-	-	-	-	-	9	4	2.232	0.558

APPENDIX A

Table A - Annual TEDE for Agreement State Licensees
2006

PROGRAM CODE - LICENSEE NAME	LICENSE#	No Meas. Exposure	Meas. <0.10	0.10-0.25	0.25-0.50	0.50-0.75	0.75-1.00	1.00-2.00	2.00-3.00	3.00-4.00	4.00-5.00	5.00-6.00	6.00-12.00	>12.0	Total Number Monitored	Number with Meas. Dose	Total Collective TEDE (person-rem)	Average Meas. TEDE (rem)
MANUFACTURING AND DISTRIBUTION – NUCLEAR PHARMACIES – 02500																		
CARDINAL HEALTH - AK	0707RA907-912	-	5	2	1	-	-	-	-	-	-	-	-	-	8	8	0.789	0.099
CARDINAL HEALTH - AL	1068	1	5	1	-	-	-	-	-	-	-	-	-	-	7	6	0.253	0.042
CARDINAL HEALTH - AL	1168	8	26	5	1	-	-	-	-	-	-	-	-	-	40	32	1.778	0.056
CARDINAL HEALTH - AR	ARK-642-02500	20	53	7	1	-	-	-	-	-	-	-	-	-	81	61	3.014	0.049
CARDINAL HEALTH - AZ	01-084	-	6	1	2	-	-	-	-	-	-	-	-	-	9	9	0.986	0.110
CARDINAL HEALTH - AZ	07-123	18	46	12	6	-	-	-	-	-	-	-	-	-	82	64	4.786	0.075
CARDINAL HEALTH - CA	2891-37	2	12	4	2	-	-	-	-	-	-	-	-	-	20	18	1.630	0.091
CARDINAL HEALTH - CA	3317-19	-	15	2	-	-	-	-	-	-	-	-	-	-	17	17	0.780	0.046
CARDINAL HEALTH - CA	3426-43	1	9	-	-	-	-	-	-	-	-	-	-	-	10	9	0.183	0.020
CARDINAL HEALTH - CA	3469-1	-	4	1	1	-	-	1	-	-	-	-	-	-	7	7	2.117	0.302
CARDINAL HEALTH - CA	3469-2	-	2	1	1	-	-	1	1	-	-	-	-	-	6	6	4.485	0.748
CARDINAL HEALTH - CA	3673-34	6	30	4	-	-	-	-	-	-	-	-	-	-	40	34	1.182	0.035
CARDINAL HEALTH - CA	3822-19	8	33	8	7	2	1	-	-	-	-	-	-	-	59	51	6.603	0.129
CARDINAL HEALTH - CA	3832-01	5	27	5	1	-	-	-	-	-	-	-	-	-	38	33	2.057	0.062
CARDINAL HEALTH - CA	4905-10	5	25	3	-	-	-	-	-	-	-	-	-	-	33	28	1.240	0.044
CARDINAL HEALTH - CA	4999-30	1	15	13	8	-	-	-	-	-	-	-	-	-	37	36	5.211	0.145
CARDINAL HEALTH - CA	5218-36	1	13	-	3	5	1	-	-	-	-	-	-	-	23	22	5.163	0.235
CARDINAL HEALTH - CA	5905-15	-	7	-	-	-	-	-	-	-	-	-	-	-	7	7	0.211	0.030
CARDINAL HEALTH - CA	5910-50	4	3	1	-	-	-	-	-	-	-	-	-	-	8	4	0.164	0.041
CARDINAL HEALTH - CA	6321-45	3	3	1	-	-	-	-	-	-	-	-	-	-	7	4	0.385	0.096
CARDINAL HEALTH - CA	6691-33	8	9	-	-	-	-	-	-	-	-	-	-	-	17	9	0.209	0.023
CARDINAL HEALTH - CA	6924-36	-	2	-	-	2	1	1	-	-	-	-	-	-	6	6	3.271	0.545
CARDINAL HEALTH - CA	6925-19	6	6	3	1	-	1	1	-	-	-	-	-	-	18	12	2.693	0.224
CARDINAL HEALTH - CO	162-06	1	9	3	3	3	1	-	-	-	-	-	-	-	20	19	5.164	0.272
CARDINAL HEALTH - CO	392-01	5	44	6	3	1	-	-	-	-	-	-	-	-	59	54	3.510	0.065
CARDINAL HEALTH - CO	392-03	5	35	7	3	1	-	-	-	-	-	-	-	-	51	46	3.383	0.074
CARDINAL HEALTH - CT	00163	13	30	4	1	-	-	-	-	-	-	-	-	-	48	35	1.839	0.053
CARDINAL HEALTH - FL	1264-9	6	15	3	5	-	-	-	-	-	-	-	-	-	29	23	2.475	0.108
CARDINAL HEALTH - FL	3273-7	1	6	-	-	-	-	-	-	-	-	-	-	-	7	6	0.120	0.020
CARDINAL HEALTH - FL	3453-1	1	12	2	1	-	-	-	-	-	-	-	-	-	16	15	0.957	0.064
CARDINAL HEALTH - FL	3453-2	17	16	4	-	-	-	-	-	-	-	-	-	-	37	20	0.757	0.038
CARDINAL HEALTH - FL	3453-3	6	13	4	-	-	-	-	-	-	-	-	-	-	23	17	0.865	0.051
CARDINAL HEALTH - FL	3453-5	5	13	3	7	-	-	-	-	-	-	-	-	-	28	23	3.345	0.145

Number of Individuals with Whole Body Doses in the Ranges (rems)

APPENDIX A

Table A - Annual TEDE for Agreement State Licensees

2006 (continued)

PROGRAM CODE - LICENSEE NAME	LICENSE#	No Meas. Exposure	Meas. <0.10	Number of Individuals with Whole Body Doses in the Ranges (rems)*											Total Number Monitored	Number with Meas. Dose	Total Collective TEDE (person-rem)	Average Meas. TEDE (rem)
				0.10-0.25	0.25-0.50	0.50-0.75	0.75-1.00	1.00-2.00	2.00-3.00	3.00-4.00	4.00-5.00	5.00-6.00	6.00-12.00	>12.0				
MANUFACTURING AND DISTRIBUTION – NUCLEAR PHARMACIES – 02500																		
CARDINAL HEALTH - FL	3453-6	5	22	8	1	-	-	1	-	-	-	-	-	-	37	32	4.047	0.126
CARDINAL HEALTH - FL	3453-7	2	19	11	5	-	-	-	-	-	-	-	-	-	37	35	4.568	0.131
CARDINAL HEALTH - FL	3453-8	7	21	6	5	-	-	-	-	-	-	-	-	-	39	32	3.335	0.104
CARDINAL HEALTH - FL	3453-9	4	15	5	-	-	-	-	-	-	-	-	-	-	24	20	1.250	0.063
CARDINAL HEALTH - FL	FL-3273-3	2	3	-	-	-	-	-	-	-	-	-	-	-	5	3	0.045	0.015
CARDINAL HEALTH - GA	GA-467-1MD	2	21	2	5	-	-	-	-	-	-	-	-	-	30	28	2.821	0.101
CARDINAL HEALTH - GA	GA-467-2MD	1	9	-	-	-	-	-	-	-	-	-	-	-	10	9	0.210	0.023
CARDINAL HEALTH - GA	GA-467-3MD	-	5	-	-	-	-	-	-	-	-	-	-	-	5	5	0.180	0.036
CARDINAL HEALTH - GA	GA-823-2MD	1	9	4	-	1	1	-	-	-	-	-	-	-	16	15	1.873	0.125
CARDINAL HEALTH - IA	0043-1-77-NP	8	13	1	-	-	-	-	-	-	-	-	-	-	22	14	0.481	0.034
CARDINAL HEALTH - IL	IL-01721-C1	8	45	10	5	-	-	-	-	-	-	-	-	-	68	60	5.084	0.085
CARDINAL HEALTH - IN	R0358-45	8	20	6	-	-	-	-	-	-	-	-	-	-	34	26	1.257	0.048
CARDINAL HEALTH - KS	20-C-495-01	8	19	3	-	-	-	-	-	-	-	-	-	-	30	22	1.140	0.052
CARDINAL HEALTH - KY	202-204-32	7	26	7	-	-	-	-	-	-	-	-	-	-	40	33	1.805	0.055
CARDINAL HEALTH - KY	202-206-32	1	27	1	-	-	-	1	-	-	-	-	-	-	30	29	1.508	0.052
CARDINAL HEALTH - KY	202-333-32	-	3	-	-	-	-	-	-	-	-	-	-	-	3	3	0.047	0.016
CARDINAL HEALTH - LA	LA-10217-L01	5	7	-	-	-	-	-	-	-	-	-	-	-	12	7	0.217	0.031
CARDINAL HEALTH - LA	LA-10336-L01	1	8	1	-	-	-	-	-	-	-	-	-	-	10	9	0.486	0.054
CARDINAL HEALTH - LA	LA-3385-L01	3	3	1	-	-	-	-	-	-	-	-	-	-	7	4	0.295	0.074
CARDINAL HEALTH - LA	LA-5115-L01	2	19	2	-	-	-	-	-	-	-	-	-	-	23	21	0.941	0.045
CARDINAL HEALTH - LA	LA-5119-L01	4	8	1	-	-	-	-	-	-	-	-	-	-	13	9	0.241	0.027
CARDINAL HEALTH - LA	LA-5394-L01	3	16	4	3	-	-	-	-	-	-	-	-	-	26	23	2.240	0.097
CARDINAL HEALTH - LA	LA-7096-L01	3	9	1	-	-	-	-	-	-	-	-	-	-	13	10	0.368	0.037
CARDINAL HEALTH - MA	4008	2	-	-	-	1	-	4	-	-	-	-	-	-	7	5	6.404	**1.281**
CARDINAL HEALTH - MA	42-0146	8	21	23	6	6	3	-	-	-	-	-	-	-	67	59	12.622	0.214
CARDINAL HEALTH - MD	MD-05-148-01	3	24	7	1	-	-	-	-	-	-	-	-	-	35	32	2.007	0.063
CARDINAL HEALTH - MD	MD-33-177-01	1	-	-	-	1	1	2	-	-	-	-	-	-	5	4	4.470	1.118
CARDINAL HEALTH - MD	MD-33-198-01	5	21	13	8	1	1	2	-	-	-	-	-	-	51	46	10.470	0.228
CARDINAL HEALTH - MI	1549-4	10	22	4	6	3	-	-	-	-	-	-	-	-	45	35	4.793	0.137
CARDINAL HEALTH - MN	1037-205-89	24	46	5	4	-	-	-	-	-	-	-	-	-	79	55	3.375	0.061
CARDINAL HEALTH - MN	1045-100-89	-	-	-	2	2	-	-	-	-	-	-	-	-	4	4	1.774	0.444
CARDINAL HEALTH - MO	2840	12	34	5	1	-	-	-	-	-	-	-	-	-	52	40	1.660	0.042
CARDINAL HEALTH - MO	6141	-	-	-	-	2	1	1	-	-	-	-	-	-	4	4	3.534	0.884

APPENDIX A

Table A - Annual TEDE for Agreement State Licensees
2006 (continued)

PROGRAM CODE - LICENSEE NAME	LICENSE#	No Meas. Exposure	Meas. <0.10	0.10-0.25	0.26-0.50	0.50-0.75	0.75-1.00	1.00-2.00	2.00-3.00	3.00-4.00	4.00-5.00	5.00-6.00	6.00-12.00	>12.0	Total Number Monitored	Number with Meas. Dose	Total Collective TEDE (person-rem)	Average Meas. TEDE (rem)
MANUFACTURING AND DISTRIBUTION – NUCLEAR PHARMACIES – 02500																		
CARDINAL HEALTH - MS	MS-493-01	1	19	2	2	2	2	-	-	-	-	-	-	-	26	25	2.708	0.108
CARDINAL HEALTH - MS	MS-781-01	-	4	3	1	1	1	-	-	-	-	-	-	-	9	9	1.882	0.209
CARDINAL HEALTH - MS	MS-924-01	3	2	2	2	1	-	-	-	-	-	-	-	-	10	7	1.518	0.217
CARDINAL HEALTH - NC	011-0794-7	4	6	1	2	-	-	-	-	-	-	-	-	-	13	9	1.113	0.124
CARDINAL HEALTH - NC	025-0794-10	2	9	3	3	2	-	-	-	-	-	-	-	-	17	15	2.949	0.197
CARDINAL HEALTH - NC	026-0794-8	3	11	4	1	-	-	-	-	-	-	-	-	-	19	16	1.413	0.088
CARDINAL HEALTH - NC	041-0794-3	2	5	-	-	-	-	-	-	-	-	-	-	-	7	5	0.117	0.023
CARDINAL HEALTH - NC	041-0794-5	3	14	1	-	-	-	-	-	-	-	-	-	-	18	15	0.467	0.031
CARDINAL HEALTH - NC	060-0794-1	7	23	4	2	1	-	-	-	-	-	-	-	-	37	30	2.203	0.073
CARDINAL HEALTH - NE	01-65-01	5	28	2	3	1	-	-	-	-	-	-	-	-	38	33	2.110	0.064
CARDINAL HEALTH - NM	RP396-02	6	10	1	1	-	-	-	-	-	-	-	-	-	18	12	0.451	0.038
CARDINAL HEALTH - NV	03-11-0150-01	5	14	4	-	-	-	-	-	-	-	-	-	-	23	18	1.032	0.057
CARDINAL HEALTH - NV	092-0794-6	7	25	10	3	3	-	-	-	-	-	-	-	-	48	41	5.065	0.124
CARDINAL HEALTH - NY	2306-3119MD	2	6	-	-	-	-	-	-	-	-	-	-	-	8	6	0.104	0.017
CARDINAL HEALTH - NY	C2328	10	8	2	1	-	-	-	-	-	-	-	-	-	21	11	0.866	0.079
CARDINAL HEALTH - NY	C2364	9	21	5	-	-	-	-	-	-	-	-	-	-	35	26	1.083	0.042
CARDINAL HEALTH - NY	C2449	3	18	3	2	-	-	-	-	-	-	-	-	-	26	23	1.312	0.057
CARDINAL HEALTH - NY	C2588	3	15	6	-	-	-	-	-	-	-	-	-	-	24	21	1.847	0.088
CARDINAL HEALTH - NY	C2593	9	21	1	-	-	-	-	-	-	-	-	-	-	31	22	0.833	0.038
CARDINAL HEALTH - NY	C2613	1	17	4	-	-	-	-	-	-	-	-	-	-	22	21	0.864	0.041
CARDINAL HEALTH - NY	C3046	5	3	-	-	-	-	-	-	-	-	-	-	-	8	3	0.116	0.039
CARDINAL HEALTH - NY	C3153	2	1	-	1	-	-	-	-	-	-	-	-	-	4	2	0.480	0.240
CARDINAL HEALTH - OH	02500180000	10	17	1	1	-	-	-	-	-	-	-	-	-	29	19	0.808	0.043
CARDINAL HEALTH - OH	02500180091	-	-	2	1	1	-	-	-	-	-	-	-	-	4	4	0.566	0.142
CARDINAL HEALTH - OH	02500250000	6	40	10	3	1	-	-	-	-	-	-	-	-	60	54	4.550	0.084
CARDINAL HEALTH - OH	02500310000	-	4	2	4	-	-	-	-	-	-	-	-	-	10	10	1.826	0.183
CARDINAL HEALTH - OH	02500490001	11	28	6	-	-	-	-	-	-	-	-	-	-	45	34	1.372	0.040
CARDINAL HEALTH - OH	02500580000	16	14	2	2	-	-	-	-	-	-	-	-	-	34	18	1.364	0.076
CARDINAL HEALTH - OH	02500770000	1	10	1	-	-	-	-	-	-	-	-	-	-	12	11	0.370	0.034
CARDINAL HEALTH - OH	02500790000	4	7	1	-	-	-	-	-	-	-	-	-	-	12	8	0.241	0.030
CARDINAL HEALTH - OH	03214250004	2	-	-	-	-	-	-	-	-	-	-	-	-	2	-	-	-
CARDINAL HEALTH - OH	03214310000	-	1	1	3	-	-	-	-	-	-	-	-	-	4	4	1.320	0.330
CARDINAL HEALTH - OH	4853-19	7	10	1	3	-	-	-	-	-	-	-	-	-	21	14	1.778	0.127

APPENDIX A

Table A - Annual TEDE for Agreement State Licensees
2006 (continued)

MANUFACTURING AND DISTRIBUTION – NUCLEAR PHARMACIES – 02500

PROGRAM CODE - LICENSEE NAME / LICENSE#	No Meas. Exposure	Meas. <0.10	0.10-0.25	0.25-0.50	0.50-0.75	0.75-1.00	1.00-2.00	2.00-3.00	3.00-4.00	4.00-5.00	5.00-6.00	6.00-12.00	>12.0	Total Number Monitored	Number with Meas. Dose	Total Collective TEDE (person-rem)	Average Meas. TEDE (rem)
CARDINAL HEALTH - OK OK-19583-02MD	-	20	4		1									25	25	1.596	0.064
CARDINAL HEALTH - OK OK-23359-02MD	8	15	1											24	16	0.555	0.035
CARDINAL HEALTH - OR ORE-90509	4	12	2											18	14	0.679	0.049
CARDINAL HEALTH - OR ORE 90703	6	12	2	3										23	17	1.673	0.098
CARDINAL HEALTH - OR ORE-90914	4	23	12	9	1	1								50	46	6.775	0.147
CARDINAL HEALTH - PA 10239-02-002	2	18	2	2										24	22	1.726	0.078
CARDINAL HEALTH - PA PA-0384	5	38	5	4	2									54	49	4.322	0.088
CARDINAL HEALTH - PA PA-0385	9	37	8	7	4									65	56	7.162	0.128
CARDINAL HEALTH - PA PA-0415	1	20	12											33	32	2.851	0.089
CARDINAL HEALTH - PA PA-0450	5	11	5											21	16	0.845	0.053
CARDINAL HEALTH - PA PA-0643	2	12	1	1										16	14	0.762	0.054
CARDINAL HEALTH - PA PA-0680	2	18	5	4	1		1							31	29	4.489	0.155
CARDINAL HEALTH - PA PA-0895	-		2	1										3	3	0.589	0.196
CARDINAL HEALTH - RI 3B-114-01	13	25												38	25	0.619	0.025
CARDINAL HEALTH - SC 448	4	16	1	3										24	20	1.521	0.076
CARDINAL HEALTH - SD 47111	1	5	10	1										17	16	1.839	0.115
CARDINAL HEALTH - TN R-19199-B15	3	26	4											33	30	2.088	0.070
CARDINAL HEALTH - TN R-33111-I14	3	22	1											26	23	0.502	0.022
CARDINAL HEALTH - TN R47080-L14	13	18	3	1										35	22	1.441	0.066
CARDINAL HEALTH - TN R-57025-C15	2	6												8	6	0.285	0.048
CARDINAL HEALTH - TN R-71029-B11	-	7												7	7	0.129	0.018
CARDINAL HEALTH - TN R-79174-D17	1	11	3	7										22	21	3.600	0.171
CARDINAL HEALTH - TX L0-1911	24	41	12	9	4	1								91	67	8.889	0.133
CARDINAL HEALTH - TX L0-1999	5	5												10	5	0.060	0.012
CARDINAL HEALTH - TX L0-2033	10	19	8	4	3									44	34	5.329	0.157
CARDINAL HEALTH - TX L0-2048	5	28	7	3										43	38	3.117	0.082
CARDINAL HEALTH - TX L0-2117	3	18	4	1										26	23	1.473	0.064
CARDINAL HEALTH - TX L0-2737	1	3												4	3	0.101	0.034
CARDINAL HEALTH - TX L0-3398	1	9												10	9	0.224	0.025
CARDINAL HEALTH - TX L0-4043	1	8	1											10	9	0.231	0.026
CARDINAL HEALTH - TX L0-4573	13	33	7	3										56	43	3.083	0.072
CARDINAL HEALTH - TX L0-4781	11	10	2											23	12	0.560	0.047
CARDINAL HEALTH - TX L0-4785	7	10	3	2										22	15	1.159	0.077

Number of Individuals with Whole Body Doses in the Ranges (rems)

APPENDIX A
Table A - Annual TEDE for Agreement State Licensees 2006 (continued)

PROGRAM CODE - LICENSEE NAME	LICENSE#	No Meas. Exposure	Meas. <0.10	Number of Individuals with Whole Body Doses in the Ranges (rems)*											Total Number Monitored	Number with Meas. Dose	Total Collective TEDE (person-rem)	Average Meas. TEDE (rem)
				0.10-0.25	0.25-0.50	0.50-0.75	0.75-1.00	1.00-2.00	2.00-3.00	3.00-4.00	4.00-5.00	5.00-6.00	6.00-12.00	>12.0				
MANUFACTURING AND DISTRIBUTION – NUCLEAR PHARMACIES – 02500																		
CARDINAL HEALTH - TX	LO-5461	1	5	1	-	-	-	-	-	-	-	-	-	-	7	6	0.222	0.037
CARDINAL HEALTH - TX	LO-5536	-	1	1	3	2	-	-	-	-	-	-	-	-	7	7	2.526	0.361
CARDINAL HEALTH - UNITED STATES	R-79272-111	-	-	-	2	-	-	-	-	-	-	-	-	-	2	2	0.767	0.384
CARDINAL HEALTH - US	065-0794-11	1	5	2	-	-	-	-	-	-	-	-	-	-	8	7	0.709	0.101
CARDINAL HEALTH - US	CH-US-001	28	50	23	21	7	3	4	2	1	-	-	-	-	139	111	34.648	0.312
CARDINAL HEALTH - UT	UT-1800225	-	13	13	4	-	-	-	-	-	-	-	-	-	26	26	2.548	0.098
CARDINAL HEALTH - VA	760-34-1	6	45	9	4	-	-	-	-	-	-	-	-	-	64	58	4.093	0.071
CARDINAL HEALTH - WA	WN-NP003-1	6	29	2	1	-	-	-	-	-	-	-	-	-	38	32	1.511	0.047
CARDINAL HEALTH - WA	WN-NP004-1	-	23	5	3	-	-	-	-	-	-	-	-	-	31	31	2.728	0.088
CARDINAL HEALTH - WA	WN-NP005-1	5	18	2	-	-	-	-	-	-	-	-	-	-	25	20	0.700	0.035
CARDINAL HEALTH - WA	WN-NP006-1	1	6	-	1	-	-	-	-	-	-	-	-	-	8	7	0.599	0.086
CARDINAL HEALTH - WA	WN-NP011-1	5	21	39	15	6	-	1	-	-	-	-	-	-	87	82	16.408	0.200
CARDINAL HEALTH - WI	025-1123-01	1	14	1	2	1	-	-	-	-	-	-	-	-	19	18	2.335	0.130
CARDINAL HEALTH - WI	087-1312-01	8	10	2	-	-	-	-	-	-	-	-	-	-	20	12	0.637	0.053
CARDINAL HEALTH - WI	141-1306-01	2	11	3	-	-	-	-	-	-	-	-	-	-	16	14	0.880	0.063
GE HEALTHCARE - ANAHEIM	4810-30	19	13	1	-	-	-	-	-	-	-	-	-	-	33	14	0.594	0.042
GE HEALTHCARE - SACRAMENTO	4809-34	23	6	2	-	-	-	-	-	-	-	-	-	-	31	8	0.696	0.087
GE HEALTHCARE - SAN DIEGO	5796-37	11	15	1	-	-	-	-	-	-	-	-	-	-	27	16	0.475	0.030
GE HEALTHCARE - SAN JOSE	4811-43	9	11	5	1	1	-	-	-	-	-	-	-	-	27	18	2.228	0.124
GE HEALTHCARE - VAN NUYS	5143-19	13	5	1	-	-	-	-	-	-	-	-	-	-	19	6	0.222	0.037
IBA MOLECULAR NORTH AMERICA, INC.	CA-7131-43	11	11	3	4	3	-	1	1	-	-	-	-	-	34	23	8.302	0.361
MALLINCKRODT, INC.	859-01	12	7	-	-	-	-	-	-	-	-	-	-	-	19	7	0.158	0.023
MORAVEK BIOCHEMICALS, INC.	2960-30	-	18	7	1	-	-	-	-	-	-	-	-	-	26	26	2.463	0.095
WYLE LABORATORIES	FL-2963-1	20	6	-	-	-	-	-	-	-	-	-	-	-	26	6	0.155	0.026
Total	**156**	**807**	**2,375**	**571**	**279**	**78**	**16**	**19**	**5**	**1**	**-**	**-**	**-**	**-**	**4,151**	**3,344**	**358.759**	**0.107**
WELL LOGGING BYPRODUCT AND/OR SNM SEALED SOURCES ONLY - 03111																		
ISOTOPE PRODUCTS LABS	1509-19	5	25	11	7	4	6	9	1	-	-	-	-	-	68	63	27.714	0.440
Total	**1**	**5**	**25**	**11**	**7**	**4**	**6**	**9**	**1**	**-**	**-**	**-**	**-**	**-**	**68**	**63**	**27.714**	**0.440**
MEASURING SYSTEMS PORTABLE GAUGES - 03121																		
CPN INTERNATIONAL, INC.	1100-07	-	4	6	5	-	-	1	-	-	-	-	-	-	16	16	4.552	0.285
Total	**1**	**-**	**4**	**6**	**5**	**-**	**-**	**1**	**-**	**-**	**-**	**-**	**-**	**-**	**16**	**16**	**4.552**	**0.285**

APPENDIX A

Table A - Annual TEDE for Agreement State Licensees
2006 (continued)

The middle block of columns (Meas. <0.10 through >12.0) falls under the spanning header **"Number of Individuals with Whole Body Doses in the Ranges (rems)"**.

PROGRAM CODE - LICENSEE NAME	LICENSE#	No Meas. Exposure	Meas. <0.10	0.10-0.25	0.25-0.50	0.50-0.75	0.75-1.00	1.00-2.00	2.00-3.00	3.00-4.00	4.00-5.00	5.00-6.00	6.00-12.00	>12.0	Total Number Monitored	Number with Meas. Dose	Total Collective TEDE (person-rem)	Average Meas. TEDE (rem)
MANUFACTURING AND DISTRIBUTION – OTHER - 03214																		
HOCHIKI AMERICA CORP	2090-30	18	-	-	-	-	-	-	-	-	-	-	-	-	18	-	-	-
J. L. SHEPHERD AND ASSOCIATES	CA 1777-19	15	3	1	3	1	-	-	-	-	-	-	-	-	23	8	2.039	0.255
Total	2	33	3	1	3	1	-	-	-	-	-	-	-	-	41	8	2.039	0.265
WASTE DISPOSAL SERVICE PROCESSING AND/OR REPACKAGING - 03234																		
ENVIRONMENTAL MANAGEMENT & CONTROLS	3546-50	-	-	2	-	2	1	-	-	-	-	-	-	-	5	5	2.601	0.520
Total	1	-	-	2	-	2	1	-	-	-	-	-	-	-	5	5	2.601	0.520
INDUSTRIAL RADIOGRAPHY – FIXED LOCATION – 03310																		
LAFAYETTE TESTING SERVICES, INC.	079-1147-01	1	7	1	-	-	-	-	-	-	-	-	-	-	9	8	0.257	0.032
METALTEK - WISCONSIN CENTRIFUGAL DIVISION	133-118--01	2	1	1	2	-	1	-	-	-	-	-	-	-	7	5	1.439	0.288
SONGS - RADIOGRAPHY GROUP	CA-5244-30	8	-	-	-	-	-	-	-	-	-	-	-	-	8	-	-	-
WAUKESHA FOUNDRY, INC. - WI	133-1337-01	4	-	-	-	-	-	-	-	-	-	-	-	-	4	-	-	-
Total	4	15	8	2	2	-	1	-	-	-	-	-	-	-	28	13	1.696	0.130
INDUSTRIAL RADIOGRAPHY – TEMPORARY JOB SITE – 03320																		
ACUREN	133-2008-01	9	6	4	10	5	4	10	3	-	-	-	-	-	51	42	32.919	0.784
ALPHA-OMEGA SERVICES, INC.	3925-19	2	8	2	3	-	-	-	-	-	-	-	-	-	15	13	1.676	0.129
AMERICAN ENGINEERING TESTING, INC.	073-2012-02	-	2	1	2	2	-	1	4	-	-	-	-	-	12	12	14.497	1.208
BRAUN INTERTEC - MN	MN1082-00-27	5	7	9	5	2	1	6	-	-	-	-	-	-	35	30	12.672	0.422
CONAM INSPECTION	IL-01225-22	3	21	16	17	4	4	12	2	-	-	-	-	-	79	76	36.716	0.483
CONAM INSPECTION	LA-10986-201	-	2	4	3	1	1	-	-	-	-	-	-	-	11	11	3.155	0.287
CONAM INSPECTION	L05010	27	27	18	14	4	5	30	9	2	-	-	-	-	136	109	84.643	0.777
CONAM INSPECTION	OK-31055-01	-	1	-	-	-	-	-	-	-	-	-	-	-	1	1	0.048	0.048
CONAM INSPECTION	090-1058-1	1	6	6	7	4	3	12	1	-	-	-	-	-	40	39	28.461	0.730
CONAM INSPECTION	1075	-	5	4	8	2	1	7	2	-	-	-	-	-	29	29	20.563	0.709
CONAM INSPECTION	16-5591	-	23	7	5	-	2	-	-	-	-	-	-	-	37	37	4.985	0.135
CONAM INSPECTION	21-B878	-	6	4	3	1	1	3	-	-	-	-	-	-	18	18	8.285	0.460
CONAM INSPECTION	3080-4315	-	2	-	1	1	1	-	-	-	-	-	-	-	5	5	2.012	0.402
CONAM INSPECTION	03320460000	2	4	1	1	1	3	1	1	-	-	-	-	-	16	14	8.097	0.578
CONAM INSPECTION	4832-19	1	1	2	9	4	6	1	1	-	-	-	-	-	10	9	6.756	0.751
CONAM INSPECTION	4886-48	11	13	14	8	3	4	7	-	-	-	-	-	-	48	37	16.987	0.459
CONAM INSPECTION	730	3	20	-	3	2	1	-	-	-	-	-	-	-	59	56	19.822	0.354
CONAM INSPECTION	963-01	-	1	1	3	2	1	1	-	-	-	-	-	-	8	8	4.107	0.513
CONAM INSPECTION	963-01	1	3	2	5	4	3	3	1	-	-	-	-	-	21	20	13.660	0.683

APPENDIX A

Table A - Annual TEDE for Agreement State Licensees

2006 (continued)

PROGRAM CODE-LICENSEE NAME	LICENSE#	No Meas. Exposure	Number of Individuals with Whole Body Doses in the Ranges (rems)*												Total Number Monitored	Number with Meas. Dose	Total Collective TEDE (person-rem)	Average Meas. TEDE (rem)
			Meas. <0.10	0.10-0.25	0.25-0.50	0.50-0.75	0.75-1.00	1.00-2.00	2.00-3.00	3.00-4.00	4.00-5.00	5.00-6.00	6.00-12.00	>12.0				
INDUSTRIAL RADIOGRAPHY – TEMPORARY JOB SITE – 03320																		
CONSTRUCTION MATERIALS TESTING, INC.	0799-07	2	1	-	2	-	-	-	-	-	-	-	-	-	6	4	1.029	0.257
EDGE INSPECTION GROUP, INC	CA-7214-48	1	6	1	-	2	1	2	-	-	-	-	-	-	12	11	5.435	0.494
GLOBE X-RAY SERVICES, INC.	OK-15194-02	2	6	3	5	4	6	5	4	2	1	-	-	-	38	36	37.428	1.040
GREAT LAKES TESTING, INC.	009-1116-01	4	5	4	2	2	1	2	-	-	-	-	-	-	20	16	6.252	0.391
INTERNATIONAL INSPECTION, INC	5046-19	4	2	3	2	-	-	-	-	-	-	-	-	-	11	7	1.436	0.205
IRIS, NDT, INC.	L-04769	4	15	8	16	2	5	10	4	-	-	-	-	-	64	60	35.410	0.590
IRIS, NDT, INC.	OK-30246-02	3	12	6	9	6	8	21	8	3	2	-	-	-	78	75	83.585	1.114
NATIONAL SERVICE DEVELOPMENT GROUP	6313-30	2	3	2	3	1	-	3	-	-	-	-	-	-	14	12	5.443	0.454
NDE SERVICES, INC.	406-01	2	1	1	2	-	1	-	-	-	-	-	-	-	7	5	1.591	0.318
NDT LABORATORIES, INC.	1651-43	2	1	3	-	-	-	-	-	-	-	-	-	-	6	4	0.524	0.131
NDT SPECIALISTS, INC.	079-1199-01	2	4	1	-	2	2	1	-	1	-	-	-	-	13	11	7.759	0.705
SUBSURFACE IMAGING, INC.	7449-19	1	1	-	1	1	-	-	1	-	-	-	-	-	5	4	3.197	0.799
TC INSPECTION, INC.	5299-07	17	13	9	10	3	4	9	2	-	-	-	-	-	67	50	28.176	0.564
TEAM INDUSTRIAL SERVICES, INC	079-2005-01	-	6	3	2	-	1	-	-	-	-	-	-	-	12	12	2.544	0.212
TEAM INDUSTRIAL SERVICES, INC	388-01	2	6	2	5	2	2	2	-	-	-	-	-	-	21	19	8.205	0.432
TESTING ENGINEERS, INC.	1898-01	-	4	1	1	-	-	1	-	-	-	-	-	-	7	7	1.912	0.273
TWIN PORTS TESTING, INC.	031-1317-02	5	6	-	2	3	-	-	-	-	-	-	-	-	16	11	2.477	0.225
VALLEY INDUSTRIAL X-RAY & INSPECTION	CA-4182-15	31	30	8	14	7	5	17	4	-	-	-	-	-	116	85	48.412	0.570
WESTERN INDUSTRIAL X-RAY	4424-48	-	-	2	1	2	-	2	1	-	-	-	-	-	8	8	7.307	0.913
WESTERN X-RAY CORP	5324-56	-	7	1	1	-	-	1	-	-	-	-	-	-	10	10	2.295	0.230
YUBA HEAT TRANSFER	OK-13735-01	1	3	-	-	-	-	-	-	-	-	-	-	-	5	4	0.344	0.086
Total	40	150	290	155	188	77	71	176	49	8	3	-	-	-	1,167	1,017	610.822	0.601
IRRADIATORS - OTHER - LESS THAN 10000 CURIES - 03511																		
INDUSTRIAL NUCLEAR CO., INC.	2229-01	6	2	-	-	-	-	1	-	-	-	-	-	-	11	5	2.474	0.495
Total	1	6	2	-	-	-	2	1	-	-	-	-	-	-	11	5	2.474	0.495

APPENDIX A

Table A - Annual TEDE for Agreement State Licensees

2007

PROGRAM CODE - LICENSEE NAME	LICENSE#	No Meas. Exposure	Meas. <0.10	0.10-0.25	0.25-0.50	0.50-0.75	0.75-1.00	1.00-2.00	2.00-3.00	3.00-4.00	4.00-5.00	5.00-6.00	6.00-12.00	>12.0	Total Number Monitored	Number with Meas. Dose	Total Collective TEDE (person-rem)	Average Meas. TEDE (rem)
MANUFACTURING AND DISTRIBUTION – NUCLEAR PHARMACIES – 02500																		
CARDINAL HEALTH - AK	0707RA907-912	-	4	1	-	-	-	-	-	-	-	-	-	-	5	5	0.211	0.042
CARDINAL HEALTH - AL	1068	1	2	2	-	-	-	-	-	-	-	-	-	-	5	4	0.338	0.085
CARDINAL HEALTH - AL	1168	5	20	8	6	2	1	-	-	-	-	-	-	-	42	37	5.576	0.151
CARDINAL HEALTH - AR	ARK-642-02500	12	57	6	-	-	-	-	-	-	-	-	-	-	75	63	2.581	0.041
CARDINAL HEALTH - AZ	01-084	1	6	-	2	-	-	-	-	-	-	-	-	-	9	8	0.782	0.098
CARDINAL HEALTH - AZ	07-123	6	39	11	5	1	-	-	-	-	-	-	-	-	62	56	4.780	0.085
CARDINAL HEALTH - CA	2891-37	3	13	4	2	-	-	-	-	-	-	-	-	-	22	19	1.854	0.098
CARDINAL HEALTH - CA	3317-1C	3	17	2	-	-	-	-	-	-	-	-	-	-	22	19	0.819	0.043
CARDINAL HEALTH - CA	3426-43	2	6	2	-	-	-	-	-	-	-	-	-	-	10	8	0.372	0.047
CARDINAL HEALTH - CA	3469-1	1	1	-	2	2	-	-	-	-	-	-	-	-	6	5	1.811	0.362
CARDINAL HEALTH - CA	3469-2	-	-	-	2	-	2	-	1	-	-	-	-	-	5	5	6.045	1.209
CARDINAL HEALTH - CA	3673-34	5	27	5	-	-	-	-	-	-	-	-	-	-	37	32	1.276	0.040
CARDINAL HEALTH - CA	3822-19	6	33	13	4	6	-	-	-	-	-	-	-	-	62	56	7.701	0.138
CARDINAL HEALTH - CA	3832-01	7	31	3	1	-	-	-	-	-	-	-	-	-	42	35	1.686	0.048
CARDINAL HEALTH - CA	4905-10	6	23	4	-	-	-	-	-	-	-	-	-	-	33	27	1.008	0.037
CARDINAL HEALTH - CA	4999-30	6	22	6	1	-	-	-	-	-	-	-	-	-	35	29	2.066	0.071
CARDINAL HEALTH - CA	5218-36	3	13	-	2	3	1	-	-	-	-	-	-	-	22	19	3.734	0.197
CARDINAL HEALTH - CA	5905-15	-	7	2	-	-	-	-	-	-	-	-	-	-	9	9	0.562	0.062
CARDINAL HEALTH - CA	5910-50	2	2	1	1	-	-	-	-	-	-	-	-	-	6	4	0.644	0.161
CARDINAL HEALTH - CA	6321-45	3	5	-	-	-	-	-	-	-	-	-	-	-	8	5	0.048	0.010
CARDINAL HEALTH - CA	6691-33	5	12	-	-	-	-	-	-	-	-	-	-	-	17	12	0.146	0.012
CARDINAL HEALTH - CA	6924-36	-	-	-	2	-	1	1	-	-	-	-	-	-	4	4	2.886	0.722
CARDINAL HEALTH - CA	6925-19	2	11	3	4	-	1	-	-	-	-	-	-	-	21	19	3.127	0.165
CARDINAL HEALTH - CO	162-06	4	8	3	7	1	-	2	-	-	-	-	-	-	25	21	5.614	0.267
CARDINAL HEALTH - CO	392-01	3	44	7	2	-	-	-	-	-	-	-	-	-	56	53	2.644	0.050
CARDINAL HEALTH - CO	392-03	1	39	7	2	-	-	-	-	-	-	-	-	-	49	48	2.455	0.051
CARDINAL HEALTH - CT	00163	5	29	8	-	-	-	-	-	-	-	-	-	-	42	37	1.792	0.048
CARDINAL HEALTH - FL	1264-9	1	12	4	5	-	-	-	-	-	-	-	-	-	22	21	2.699	0.129
CARDINAL HEALTH - FL	3273-7	-	6	1	-	-	-	-	-	-	-	-	-	-	7	7	0.190	0.027
CARDINAL HEALTH - FL	3453-1	4	23	9	-	-	-	-	-	-	-	-	-	-	36	32	1.787	0.056
CARDINAL HEALTH - FL	3453-2	11	18	5	-	-	-	-	-	-	-	-	-	-	34	23	1.003	0.044
CARDINAL HEALTH - FL	3453-3	4	21	2	-	-	-	-	-	-	-	-	-	-	27	23	0.736	0.032
CARDINAL HEALTH - FL	3453-5	5	18	3	3	1	-	-	-	-	-	-	-	-	30	25	3.001	0.120

APPENDIX A

Table A - Annual TEDE for Agreement State Licensees

2007 (continued)

PROGRAM CODE - LICENSEE NAME	LICENSE#	No Meas. Exposure	Meas. <0.10	0.10-0.25	0.25-0.50	0.50-0.75	0.75-1.00	1.00-2.00	2.00-3.00	3.00-4.00	4.00-5.00	5.00-6.00	6.00-12.00	>12.0	Total Number Monitored	Number with Meas. Dose	Total Collective TEDE (person-rem)	Average Meas. TEDE (rem)
MANUFACTURING AND DISTRIBUTION – NUCLEAR PHARMACIES – 02500																		
CARDINAL HEALTH - FL	3453-6	11	20	6	-	-	-	-	-	-	-	-	-	-	37	26	1.439	0.055
CARDINAL HEALTH - FL	3453-7	6	21	8	5	-	-	-	-	-	-	-	-	-	40	34	3.855	0.113
CARDINAL HEALTH - FL	3453-8	10	25	4	1	1	-	-	-	-	-	-	-	-	41	31	2.277	0.073
CARDINAL HEALTH - FL	3453-9	2	19	4	-	-	-	-	-	-	-	-	-	-	25	23	1.055	0.046
CARDINAL HEALTH - FL	FL-3273-3	2	-	-	-	-	-	-	-	-	-	-	-	-	2	-	-	-
CARDINAL HEALTH - GA	GA-467-1MD	5	14	4	3	-	-	-	-	-	-	-	-	-	26	21	2.146	0.102
CARDINAL HEALTH - GA	GA-467-2MD	2	9	2	-	-	-	-	-	-	-	-	-	-	13	11	0.479	0.044
CARDINAL HEALTH - GA	GA-467-3MD	-	5	-	-	-	-	-	-	-	-	-	-	-	5	5	0.051	0.010
CARDINAL HEALTH - GA	GA-823-2MD	-	9	4	1	-	-	-	-	-	-	-	-	-	14	14	1.363	0.097
CARDINAL HEALTH - IA	0043-1-77-NP	4	24	-	-	-	-	-	-	-	-	-	-	-	28	24	0.638	0.027
CARDINAL HEALTH - IL	IL-01721-01	12	47	5	2	-	-	-	-	-	-	-	-	-	66	54	3.003	0.056
CARDINAL HEALTH - IN	R0358-45	15	19	3	-	-	-	-	-	-	-	-	-	-	37	22	0.759	0.035
CARDINAL HEALTH - KS	20-C-495-01	2	19	5	1	-	-	-	-	-	-	-	-	-	27	25	1.603	0.064
CARDINAL HEALTH - KY	202-204-32	3	37	3	-	-	-	-	-	-	-	-	-	-	43	40	1.508	0.038
CARDINAL HEALTH - KY	202-206-32	2	29	2	-	-	-	-	-	-	-	-	-	-	33	31	0.963	0.031
CARDINAL HEALTH - KY	202-333-32	-	3	-	-	-	-	-	-	-	-	-	-	-	3	3	0.031	0.010
CARDINAL HEALTH - LA	LA-10217-L01	4	6	-	-	-	-	-	-	-	-	-	-	-	10	6	0.144	0.024
CARDINAL HEALTH - LA	LA-10336-L01	2	8	-	-	-	-	-	-	-	-	-	-	-	10	8	0.227	0.028
CARDINAL HEALTH - LA	LA-3385-L01	-	18	1	1	-	-	-	-	-	-	-	-	-	20	20	0.898	0.045
CARDINAL HEALTH - LA	LA-5115-L01	-	18	1	-	-	-	-	-	-	-	-	-	-	19	19	0.960	0.051
CARDINAL HEALTH - LA	LA-5119-L01	1	10	-	-	-	-	-	-	-	-	-	-	-	11	10	0.185	0.019
CARDINAL HEALTH - LA	LA-5394-L01	7	16	4	1	1	-	-	-	-	-	-	-	-	29	22	1.871	0.085
CARDINAL HEALTH - LA	LA-7096-L01	4	9	-	1	-	-	-	-	-	-	-	-	-	14	10	0.465	0.047
CARDINAL HEALTH - MA	4008	-	1	1	-	-	1	-	2	2	-	-	-	-	7	7	12.552	1.793
CARDINAL HEALTH - MA	42-0146	9	24	17	6	3	4	1	-	-	-	-	-	-	64	55	12.763	0.232
CARDINAL HEALTH - MD	MD-05-148-01	1	25	3	4	-	-	-	-	-	-	-	-	-	33	32	2.745	0.086
CARDINAL HEALTH - MD	MD-33-177-01	-	-	-	1	1	1	1	-	-	-	-	-	-	4	4	3.323	0.831
CARDINAL HEALTH - MD	MD-33-198-01	4	22	11	3	1	1	2	-	-	-	-	-	-	44	40	10.074	0.252
CARDINAL HEALTH - ME	19233	1	-	-	-	-	-	-	-	-	-	-	-	-	1	-	-	-
CARDINAL HEALTH - MI	1549-4	7	20	5	1	1	-	-	-	-	-	-	-	-	34	27	2.135	0.079
CARDINAL HEALTH - MN	1037-205-89	16	49	5	3	-	-	-	-	-	-	-	-	-	73	57	2.917	0.051
CARDINAL HEALTH - MN	1045-100-89	-	-	-	-	3	1	-	-	-	-	-	-	-	4	4	2.500	0.625
CARDINAL HEALTH - MO	2840	13	39	4	1	-	-	-	-	-	-	-	-	-	57	44	1.685	0.038

Number of Individuals with Whole Body Doses in the Ranges (rems)*

APPENDIX A

Table A - Annual TEDE for Agreement State Licensees
2007 (continued)

PROGRAM CODE - LICENSEE NAME	LICENSE#	No Meas. Exposure	Meas. <0.10	0.10-0.25	0.25-0.50	0.50-0.75	0.75-1.00	1.00-2.00	2.00-3.00	3.00-4.00	4.00-5.00	5.00-6.00	6.00-12.00	>12.0	Total Number Monitored	Number with Meas. Dose	Total Collective TEDE (person-rem)	Average Meas. TEDE (rem)
MANUFACTURING AND DISTRIBUTION – NUCLEAR PHARMACIES – 02500																		
CARDINAL HEALTH - MO	6141	-	1	-	-	-	2	2	-	-	-	-	-	-	5	5	4.595	0.919
CARDINAL HEALTH - MS	MS-493-01	2	11	4	4	-	1	-	-	-	-	-	-	-	22	20	3.368	0.168
CARDINAL HEALTH - MS	MS-781-01	2	3	3	-	-	1	-	-	-	-	-	-	-	9	7	1.428	0.204
CARDINAL HEALTH - MS	MS-924-01	2	4	3	1	-	-	-	-	-	-	-	-	-	10	8	1.091	0.136
CARDINAL HEALTH - MS	MS-974-01	-	-	-	1	-	1	1	-	-	-	-	-	-	3	3	2.539	0.846
CARDINAL HEALTH - NC	011-0794-7	3	5	2	-	1	-	-	-	-	-	-	-	-	11	8	1.095	0.137
CARDINAL HEALTH - NC	025-0794-10	2	10	2	2	-	-	-	-	-	-	-	-	-	16	14	1.232	0.088
CARDINAL HEALTH - NC	026-0794-8	4	13	1	-	-	-	-	-	-	-	-	-	-	18	14	0.680	0.049
CARDINAL HEALTH - NC	041-0794-5	3	16	3	-	-	-	-	-	-	-	-	-	-	22	19	0.693	0.036
CARDINAL HEALTH - NC	060-0794-1	5	28	5	3	1	1	-	-	-	-	-	-	-	43	38	4.025	0.106
CARDINAL HEALTH - NE	01-65-01	6	31	6	-	-	-	-	-	-	-	-	-	-	43	37	2.097	0.057
CARDINAL HEALTH - NM	RP396-02	7	11	1	-	-	-	-	-	-	-	-	-	-	19	12	0.269	0.022
CARDINAL HEALTH - NV	03-11-0150-01	5	14	4	-	-	-	-	-	-	-	-	-	-	23	18	1.101	0.061
CARDINAL HEALTH - NV	092-0794-8	2	29	7	4	1	-	-	-	-	-	-	-	-	43	41	4.315	0.105
CARDINAL HEALTH - NY	2306-3119MD	3	1	-	-	-	-	-	-	-	-	-	-	-	4	1	0.004	0.004
CARDINAL HEALTH - NY	C2328	5	12	4	2	-	-	-	-	-	-	-	-	-	23	18	1.719	0.096
CARDINAL HEALTH - NY	C2364	19	19	3	-	-	-	-	-	-	-	-	-	-	41	22	0.933	0.042
CARDINAL HEALTH - NY	C2449	13	8	2	2	-	-	-	-	-	-	-	-	-	25	12	1.034	0.086
CARDINAL HEALTH - NY	C2588	5	14	1	-	-	-	-	-	-	-	-	-	-	20	15	0.548	0.037
CARDINAL HEALTH - NY	C2593	8	23	2	-	-	-	-	-	-	-	-	-	-	33	25	0.766	0.031
CARDINAL HEALTH - NY	C2613	3	19	2	-	-	-	-	-	-	-	-	-	-	24	21	0.700	0.033
CARDINAL HEALTH - NY	C3046	4	14	-	-	-	-	-	-	-	-	-	-	-	18	14	0.203	0.015
CARDINAL HEALTH - NY	C3153	2	2	1	2	1	-	-	-	-	-	-	-	-	8	6	1.671	0.279
CARDINAL HEALTH - OH	02500180000	13	17	2	-	-	-	-	-	-	-	-	-	-	32	19	0.732	0.039
CARDINAL HEALTH - OH	02500180091	-	1	1	2	1	-	-	-	-	-	-	-	-	5	5	1.675	0.335
CARDINAL HEALTH - OH	02500250000	19	30	10	4	1	-	-	-	-	-	-	-	-	64	45	4.403	0.098
CARDINAL HEALTH - OH	02500310000	-	5	5	2	2	-	-	-	-	-	-	-	-	14	14	2.733	0.195
CARDINAL HEALTH - OH	02500490001	17	26	5	-	-	-	-	-	-	-	-	-	-	48	31	1.320	0.043
CARDINAL HEALTH - OH	02500580000	11	21	3	3	-	-	-	-	-	-	-	-	-	38	27	1.867	0.069
CARDINAL HEALTH - OH	02500770000	5	5	-	-	-	-	-	-	-	-	-	-	-	10	5	0.044	0.009
CARDINAL HEALTH - OH	02500790000	-	12	1	-	-	-	-	-	-	-	-	-	-	13	13	0.321	0.025
CARDINAL HEALTH - OH	0321431D000	-	-	-	1	3	-	-	-	-	-	-	-	-	4	4	1.948	0.487
CARDINAL HEALTH - OH	4853-19	8	11	1	2	-	-	-	-	-	-	-	-	-	22	14	1.103	0.079

APPENDIX A
Table A - Annual TEDE for Agreement State Licensees
2007 (continued)

PROGRAM CODE - LICENSEE NAME	LICENSE#	No Meas. Exposure	Number of Individuals with Whole Body Doses in the Ranges (rems)*												Total Number Monitored	Number with Meas. Dose	Total Collective TEDE (person-rem)	Average Meas. TEDE (rem)
			Meas. <0.10	0.10-0.25	0.25-0.50	0.50-0.75	0.75-1.00	1.00-2.00	2.00-3.00	3.00-4.00	4.00-5.00	5.00-6.00	6.00-12.00	>12.0				
MANUFACTURING AND DISTRIBUTION – NUCLEAR PHARMACIES – 02500																		
CARDINAL HEALTH - OK	OK-19583-02MD	2	15	5	1	-	-	-	-	-	-	-	-	-	23	21	1.466	0.070
CARDINAL HEALTH - OK	OK-23359-02MD	5	17	2	1	-	-	-	-	-	-	-	-	-	25	20	1.053	0.053
CARDINAL HEALTH - OR	ORE-90509	2	12	4	1	-	-	-	-	-	-	-	-	-	19	17	1.264	0.074
CARDINAL HEALTH - OR	ORE-90703	1	12	1	1	2	1	-	-	-	-	-	-	-	17	16	1.948	0.122
CARDINAL HEALTH - OR	ORE-90914	4	16	11	9	1	1	1	-	-	-	-	-	-	43	39	7.764	0.199
CARDINAL HEALTH - PA	10239-02-002	5	14	5	4	1	2	-	-	-	-	-	-	-	30	25	4.045	0.162
CARDINAL HEALTH - PA	AC50-58988	-	2	1	-	-	-	-	-	-	-	-	-	-	3	3	0.521	0.174
CARDINAL HEALTH - PA	PA-0384	10	40	8	4	1	1	-	-	-	-	-	-	-	64	54	5.123	0.095
CARDINAL HEALTH - PA	PA-0385	7	35	10	7	4	-	-	-	-	-	-	-	-	63	56	7.630	0.136
CARDINAL HEALTH - PA	PA-0415	-	20	9	-	-	-	-	-	-	-	-	-	-	29	29	2.518	0.087
CARDINAL HEALTH - PA	PA-0460	7	13	4	-	-	-	-	-	-	-	-	-	-	24	17	1.038	0.061
CARDINAL HEALTH - PA	PA-0643	4	11	1	-	-	-	-	-	-	-	-	-	-	16	12	0.460	0.038
CARDINAL HEALTH - PA	PA-0680	6	19	3	3	2	-	-	-	-	-	-	-	-	33	27	3.201	0.119
CARDINAL HEALTH - RI	3B-114-01	9	28	1	-	-	-	-	-	-	-	-	-	-	38	29	0.668	0.023
CARDINAL HEALTH - SC	448	3	15	2	3	-	-	-	-	-	-	-	-	-	23	20	1.629	0.081
CARDINAL HEALTH - SD	47111	2	14	2	-	-	-	-	-	-	-	-	-	-	18	16	1.151	0.072
CARDINAL HEALTH - TN	R-19199-B15	8	24	2	-	-	-	-	-	-	-	-	-	-	34	26	0.947	0.036
CARDINAL HEALTH - TN	R-33111-I14	2	17	2	-	-	-	-	-	-	-	-	-	-	21	19	0.643	0.034
CARDINAL HEALTH - TN	R47080-L14	5	27	5	1	-	-	-	-	-	-	-	-	-	38	33	1.495	0.045
CARDINAL HEALTH - TN	R-57025-C15	1	5	1	-	-	-	-	-	-	-	-	-	-	7	6	0.198	0.033
CARDINAL HEALTH - TN	R-71029-B11	1	7	-	-	-	-	-	-	-	-	-	-	-	8	7	0.118	0.017
CARDINAL HEALTH - TN	R-79174-D17	3	11	5	6	3	-	-	-	-	-	-	-	-	25	22	3.003	0.137
CARDINAL HEALTH - TX	L0-1911	21	37	8	10	3	-	-	-	-	-	-	-	-	79	58	6.874	0.119
CARDINAL HEALTH - TX	L0-1999	-	6	-	-	-	-	-	-	-	-	-	-	-	6	6	0.125	0.021
CARDINAL HEALTH - TX	L0-2033	12	19	9	5	3	-	-	-	-	-	-	-	-	48	36	5.352	0.149
CARDINAL HEALTH - TX	L0-2048	14	21	6	3	1	-	-	-	-	-	-	-	-	45	31	2.972	0.096
CARDINAL HEALTH - TX	L0-2117	1	19	2	1	-	-	-	-	-	-	-	-	-	23	22	0.992	0.045
CARDINAL HEALTH - TX	L0-2737	2	3	-	-	-	-	-	-	-	-	-	-	-	5	3	0.066	0.022
CARDINAL HEALTH - TX	L0-3398	3	7	-	-	-	-	-	-	-	-	-	-	-	10	7	0.220	0.031
CARDINAL HEALTH - TX	L0-4043	2	6	-	-	-	-	-	-	-	-	-	-	-	8	6	0.126	0.021
CARDINAL HEALTH - TX	L0-4573	12	26	10	5	1	-	-	-	-	-	-	-	-	54	42	4.843	0.115
CARDINAL HEALTH - TX	L0-4781	12	11	1	-	-	-	-	-	-	-	-	-	-	24	12	0.518	0.043
CARDINAL HEALTH - TX	L0-4785	3	11	4	-	-	-	-	-	-	-	-	-	-	18	15	0.766	0.051

APPENDIX A

Table A - Annual TEDE for Agreement State Licensees

2007 (continued)

PROGRAM CODE - LICENSEE NAME	LICENSE#	No Meas. Exposure	Meas. <0.10	Number of Individuals with Whole Body Doses in the Ranges (rems)*											Total Number Monitored	Number with Meas. Dose	Total Collective TEDE (person-rem)	Average Meas. TEDE (rem)
				0.10-0.25	0.25-0.50	0.50-0.75	0.75-1.00	1.00-2.00	2.00-3.00	3.00-4.00	4.00-5.00	5.00-6.00	6.00-12.00	>12.0				
MANUFACTURING AND DISTRIBUTION – NUCLEAR PHARMACIES – 02500																		
CARDINAL HEALTH - TX	LO-5461	1	6	-	-	-	-	-	-	-	-	-	-	-	7	6	0.120	0.020
CARDINAL HEALTH - TX	LO-5536	-	-	1	1	2	-	-	-	-	-	-	-	-	4	4	1.638	0.410
CARDINAL HEALTH - TX	LO-5610	1	1	1	-	1	-	-	-	-	-	-	-	-	5	4	3.550	0.888
CARDINAL HEALTH - UNITED STATES	R-79272-111	1	-	-	-	2	1	-	-	-	-	-	-	-	4	3	2.204	0.735
CARDINAL HEALTH - US	065-0794-11	-	6	2	1	-	-	-	-	-	-	-	-	-	9	9	0.714	0.079
CARDINAL HEALTH - US	CH-US-001	20	40	16	10	4	3	1	1	-	-	-	-	-	95	75	17.850	0.238
CARDINAL HEALTH - UT	UT-180C225	-	18	7	1	-	-	-	-	-	-	-	-	-	26	26	2.280	0.088
CARDINAL HEALTH - VA	760-34-1	7	46	9	1	-	-	-	-	-	-	-	-	-	63	56	3.246	0.058
CARDINAL HEALTH - WA	WN-NP003-1	3	28	1	2	-	-	-	-	-	-	-	-	-	34	31	1.625	0.052
CARDINAL HEALTH - WA	WN-NP004-1	7	21	5	1	-	-	-	-	-	-	-	-	-	34	27	1.589	0.059
CARDINAL HEALTH - WA	WN-NP005-1	6	18	2	-	-	-	-	-	-	-	-	-	-	26	20	0.661	0.033
CARDINAL HEALTH - WA	WN-NP006-1	2	4	2	-	-	-	-	-	-	-	-	-	-	8	6	0.473	0.079
CARDINAL HEALTH - WA	WN-NP011-1	17	49	11	13	1	1	1	-	-	-	1	-	-	93	76	15.925	0.210
CARDINAL HEALTH - WI	025-1125-01	3	14	4	-	-	1	-	-	-	-	-	-	-	22	19	1.926	0.101
CARDINAL HEALTH - WI	087-1312-01	11	9	3	-	-	1	-	-	-	-	-	-	-	24	13	1.648	0.127
CARDINAL HEALTH - WI	141-1306-01	-	14	-	-	-	-	-	-	-	-	-	-	-	14	14	0.479	0.034
GE HEALTHCARE - ANAHEIM	4810-30	15	16	1	-	-	-	-	-	-	-	-	-	-	32	17	0.599	0.035
GE HEALTHCARE - SACRAMENTO	4809-34	30	3	3	-	-	-	-	-	-	-	-	-	-	36	6	0.426	0.071
GE HEALTHCARE - SAN DIEGO	5796-37	19	10	-	-	-	-	-	-	-	-	-	-	-	29	10	0.364	0.036
GE HEALTHCARE - SAN JOSE	4811-43	8	12	3	1	1	-	-	-	-	-	-	-	-	25	17	2.024	0.119
GE HEALTHCARE - VAN NUYS	5143-19	14	6	-	-	-	-	-	-	-	-	-	-	-	20	6	0.150	0.025
IBA MOLECULAR NORTH AMERICA, INC.	CA-7131-43	6	20	3	3	4	1	2	1	-	-	-	-	-	40	34	12.025	0.354
MALLINCKRODT, INC.	859-01	10	7	-	-	-	-	-	-	-	-	-	-	-	17	7	0.178	0.025
MORAVEK BIOCHEMICALS, INC.	2960-30	1	22	5	-	-	-	1	-	-	-	-	-	-	29	28	3.784	0.135
WYLE LABORATORIES	FL-2953-1	27	3	-	-	-	-	-	-	-	-	-	-	-	30	3	0.048	0.016
Total	157	818	2,463	510	221	72	31	14	8	3	-	-	1	-	4,141	3,323	347.103	0.104
WELL LOGGING BYPRODUCT AND/OR SNM SEALED SOURCES ONLY - 03111																		
ISOTOPE PRODUCTS LABS	1509-19	7	25	10	7	4	6	8	1	-	-	-	-	-	68	61	26.623	0.436
Total	1	7	25	10	7	4	6	8	1	-	-	-	-	-	68	61	26.623	0.436
MEASURING SYSTEMS PORTABLE GAUGES - 03121																		
CPN INTERNATIONAL, INC	1100-07	2	2	5	2	-	-	-	-	-	-	-	-	-	11	9	1.831	0.285
Total	1	2	2	5	2	-	-	-	-	-	-	-	-	-	11	9	1.831	0.285

APPENDIX A
Table A - Annual TEDE for Agreement State Licensees
2007 (continued)

PROGRAM CODE - LICENSEE NAME	LICENSE#	Number of Individuals with Whole Body Doses in the Ranges (rems)*													Total Number Monitored	Number with Meas. Dose	Total Collective TEDE (person-rem)	Average Meas. TEDE (rem)
		No Meas. Exposure	Meas. <0.10	0.10-0.25	0.25-0.50	0.50-0.75	0.75-1.00	1.00-2.00	2.00-3.00	3.00-4.00	4.00-5.00	5.00-6.00	6.00-12.00	>12.0				
MANUFACTURING AND DISTRIBUTION – OTHER - 03214																		
HOCHIKI AMERICA CORPORATION	2090-30	16													16			
J.L. SHEPHERD AND ASSOCIATES	CA 1777-19	14	2		4	1	1								22	8	2.774	0.347
Total	2	30	2		4	1	1								38	8	2.774	0.347
WASTE DISPOSAL SERVICE PROCESSING AND/OR REPACKAGING - 03234																		
ENVIRONMENTAL MANAGEMENT & CONTROLS	3546-50	1		1	1	1		1							5	4	2.141	0.535
THOMAS GRAY & ASSOCIATES	CA-2105-30	5	3												8	3	0.122	0.041
Total	2	6	3	1	1	1		1							13	7	2.263	0.323
INDUSTRIAL RADIOGRAPHY – FIXED LOCATION – 03310																		
LAFAYETTE TESTING SERVICES, INC.	079-1147-01	3	2												5	2	0.107	0.054
METALTEK - WISCONSIN CENTRIFUGAL DIVISION	133-1181-01	2	6	1	3	1									13	11	1.881	0.171
SONGS - RADIOGRAPHY GROUP	CA-5244-30	10													10			
WAUKESHA FOUNDRY, INC. - WI	133-1337-01	3													3			
Total	4	18	8	1	3	1									31	13	1.988	0.153
INDUSTRIAL RADIOGRAPHY – TEMPORARY JOB SITE – 03320																		
ACUREN	133-2008-01	7	12	7	11	9	6	12	9	1	1				75	68	64.694	0.951
ALPHA-OMEGA SERVICES, INC.	3925-19	6	6	2	2										16	10	1.168	0.117
AMERICAN ENGINEERING TESTING INC.	073-2012-02		2	2	1	1		3	2	1					12	12	14.078	1.173
BRAUN INTERTEC - MN	1082-100-27	4	9	5	7	5	2	6	3	1					42	38	26.562	0.699
CONAM INSPECTION	033204600000	4	3	2	1		3	1							14	10	5.027	0.503
CONAM INSPECTION	1517-1		1												1	1	0.014	0.014
CONAM INSPECTION	L05010	22	43	24	23	16	11	16	5						160	138	66.907	0.485
CONAM INSPECTION	WN-IR011-1	1	7	5	6	3	2	3	1	2					30	29	20.116	0.694
CONAM INSPECTION - AL	AL-1075	1	6	4	5		3	7	2						28	27	17.761	0.658
CONAM INSPECTION - CA	CA-4832-19	10	7	9	5	6	9	5	1						52	42	23.787	0.566
CONAM INSPECTION - CA	CA-4886-48	3	20	5	6	8	4	8							54	51	23.180	0.455
CONAM INSPECTION - CO	CO-963-01	1	4	3	4	5	4	13	2						36	35	31.928	0.912
CONAM INSPECTION - IL	IL-01225-22	2	21	13	9	10	3	4							62	60	19.500	0.325

APPENDIX A

Table A - Annual TEDE for Agreement State Licensees

2007 (continued)

Columns under group heading *Number of Individuals with Whole Body Doses in the Ranges (rems)**

PROGRAM CODE - LICENSEE NAME	LICENSE#	No Meas. Exposure	Meas. <0.10	0.10-0.25	0.25-0.50	0.50-0.75	0.75-1.00	1.00-2.00	2.00-3.00	3.00-4.00	4.00-5.00	5.00-6.00	6.00-12.00	>12.0	Total Number Monitored	Number with Meas. Dose	Total Collective TEDE (person-rem)	Average Meas. TEDE (rem)
CONAM INSPECTION - KY	KY-201-699-05	1	8	10	11	6	5	4	-	-	-	-	-	-	45	44	19.756	0.449
CONAM INSPECTION - LA	LA-109E6-L01	1	7	3	4	1	3	1	-	-	-	-	-	-	20	19	6.413	0.338
CONAM INSPECTION - MA	MA-16-E591	-	31	13	11	10	6	6	3	-	-	-	-	-	80	80	33.050	0.413
CONAM INSPECTION - NC	NC-090-1058-1	-	8	6	6	1	7	10	1	-	-	-	-	-	39	39	26.868	0.689
CONAM INSPECTION - NY	NY-3080-4315	2	3	1	1	-	1	2	1	1	-	-	-	-	13	11	10.718	0.974
CONAM INSPECTION - SC	SC-730	-	1	1	1	-	2	2	-	-	-	-	-	-	6	6	4.754	0.792
CONAM INSPECTION - UT	UT-180C-485	1	5	4	1	1	-	3	1	-	-	-	-	-	18	17	9.669	0.569
CONSTRUCTION MATERIALS TESTING, INC.	0799-07	1	2	1	1	1	-	-	-	-	-	-	-	-	6	5	1.111	0.222
COOPERHEAT-MQS INSPECTION	388-01	10	4	6	5	7	3	8	-	-	1	-	-	-	44	34	25.890	0.761
EDGE INSPECTION GROUP, INC.	CA-7214-48	-	3	3	3	2	1	3	2	-	-	-	-	-	17	17	12.488	0.735
GREAT LAKES TESTING, INC.	009-1116-01	3	1	6	2	1	2	3	-	-	-	-	-	-	18	15	7.318	0.488
INTERNATIONAL INSPECTION, INC.	5046-19	2	8	3	-	-	-	-	-	-	-	-	-	-	13	11	0.879	0.080
NASSCO	CA-0684-37	3	6	1	-	-	-	-	-	-	-	-	-	-	9	6	0.086	0.014
NDT LABORATORIES, INC.	1651-43	-	3	1	2	-	-	-	-	-	-	-	-	-	6	6	0.940	0.157
NDT SPECIALISTS, INC.	079-1195-01	-	3	3	1	1	1	2	-	1	-	-	-	-	12	12	7.711	0.643
QUALITY INSPECTION SERVICES, INC.	NYS C2514	-	5	4	3	-	4	7	-	-	-	-	-	-	23	23	15.611	0.679
SUBSURFACE IMAGING, INC.	7449-19	1	5	2	2	-	-	-	-	-	-	-	-	-	10	9	1.102	0.122
TC INSPECTION, INC.	5299-01	11	24	12	11	8	3	11	1	-	-	-	-	-	81	70	32.732	0.468
TECH CORR INSPECTION & ENGINEERING	L05972 TK	1	21	21	12	10	15	17	4	-	-	-	-	-	101	100	60.733	0.607
TEAM INDUSTRIAL SERVICES, INC.	079-2005-01	2	7	2	2	1	-	2	1	-	-	-	-	-	17	15	7.674	0.512
TESTING ENGINEERS, INC	1898-01	-	1	1	1	1	-	1	-	-	-	-	-	-	4	4	2.132	0.533
TWIN PORTS TESTING, INC	031-1317-02	5	3	2	2	2	-	1	1	-	-	-	-	-	16	11	5.695	0.518
VALLEY INDUSTRIAL X-RAY & INSPECTION	CA-4182-- 5	34	28	16	16	7	11	18	3	2	-	-	-	-	135	101	64.178	0.635
WELDSONIX, INC.	TX-L05718	3	13	21	31	18	14	10	1	-	-	-	-	-	111	108	55.585	0.515
WESTERN INDUSTRIAL X-RAY	4424-48	-	-	1	2	1	1	2	-	-	-	-	-	-	7	7	5.633	0.805
WESTERN X-RAY CORPORATION	5324-56	3	-	-	3	-	-	1	-	-	-	-	-	-	9	6	2.712	0.452
YUBA HEAT TRANSFER	OK-13735-01	1	2	1	-	-	-	-	-	-	-	-	-	-	4	3	0.304	0.101
Total	**40**	**146**	**345**	**225**	**212**	**145**	**126**	**192**	**44**	**9**	**2**	-	-	-	**1,446**	**1,300**	**736.464**	**0.567**
IRRADIATORS - OTHER - LESS THAN 10000 CURIES - 03511																		
INDUSTRIAL NUCLEAR CO. INC	2229-01	6	-	1	2	-	1	-	-	-	-	-	-	-	10	4	2.357	0.589
Total	**1**	**6**	-	**1**	**2**	-	**1**	-	-	-	-	-	-	-	**10**	**4**	**2.357**	**0.589**

APPENDIX A
Table A - Annual TEDE for Agreement State Licensees
2008

MANUFACTURING AND DISTRIBUTION – NUCLEAR PHARMACIES – 02500

PROGRAM CODE-LICENSEE NAME	LICENSE#	No Meas. Exposure	Meas. <0.10	0.10-0.25	0.25-0.50	0.50-0.75	0.75-1.00	1.00-2.00	2.00-3.00	3.00-4.00	4.00-5.00	5.00-6.00	6.00-12.00	>12.0	Total Number Monitored	Number with Meas. Dose	Total Collective TEDE (person-rem)	Average Meas. TEDE (rem)
CARDINAL HEALTH - AK	0707RA907-912	1	2	1	-	-	-	-	-	-	-	-	-	-	4	3	0.186	0.062
CARDINAL HEALTH - AL	1068	3	3	1	-	-	-	-	-	-	-	-	-	-	7	4	0.261	0.065
CARDINAL HEALTH - AL	1168	3	27	6	3	1	-	-	-	-	-	-	-	-	40	37	3.248	0.088
CARDINAL HEALTH - AR	ARK-642-02500	16	62	2	2	-	-	-	-	-	-	-	-	-	82	66	2.507	0.038
CARDINAL HEALTH - AZ	01-084	4	7	1	1	-	-	-	-	-	-	-	-	-	13	9	0.669	0.074
CARDINAL HEALTH - AZ	07-123	15	37	14	6	-	-	-	-	-	-	-	-	-	72	57	5.560	0.098
CARDINAL HEALTH - CA	2891-37	1	17	4	2	-	-	-	-	-	-	-	-	-	24	23	1.891	0.082
CARDINAL HEALTH - CA	3317-19	2	16	-	-	-	-	-	-	-	-	-	-	-	18	16	0.423	0.026
CARDINAL HEALTH - CA	3426-43	1	8	2	-	-	-	-	-	-	-	-	-	-	11	10	0.370	0.037
CARDINAL HEALTH - CA	3469-1	-	3	-	2	-	-	1	-	-	-	-	-	-	6	6	2.030	0.338
CARDINAL HEALTH - CA	3469-2	-	1	1	-	2	1	1	-	-	-	-	-	-	6	6	4.016	0.669
CARDINAL HEALTH - CA	3673-34	3	31	4	1	-	-	-	-	-	-	-	-	-	39	36	1.646	0.046
CARDINAL HEALTH - CA	3822-19	4	34	12	5	1	1	-	-	-	-	-	-	-	57	53	6.529	0.123
CARDINAL HEALTH - CA	3832-01	4	25	6	-	1	-	-	-	-	-	-	-	-	36	32	2.195	0.069
CARDINAL HEALTH - CA	4905-10	4	25	4	-	-	-	-	-	-	-	-	-	-	33	29	1.165	0.040
CARDINAL HEALTH - CA	4999-30	5	21	7	1	-	-	-	-	-	-	-	-	-	34	29	1.800	0.062
CARDINAL HEALTH - CA	5218-36	6	7	2	2	2	2	-	-	-	-	-	-	-	21	15	3.978	0.265
CARDINAL HEALTH - CA	5905-15	-	8	-	-	-	-	-	-	-	-	-	-	-	8	8	0.295	0.037
CARDINAL HEALTH - CA	5910-50	2	4	1	-	-	-	-	-	-	-	-	-	-	7	5	0.209	0.042
CARDINAL HEALTH - CA	6321-45	3	6	-	-	-	-	-	-	-	-	-	-	-	9	6	0.172	0.029
CARDINAL HEALTH - CA	6691-33	3	10	-	-	-	-	-	-	-	-	-	-	-	13	10	0.136	0.014
CARDINAL HEALTH - CA	6924-36	1	1	1	2	1	-	-	-	-	-	-	-	-	6	5	1.537	0.307
CARDINAL HEALTH - CA	6925-19	2	11	8	2	1	1	-	-	-	-	-	-	-	25	23	3.801	0.165
CARDINAL HEALTH - CO	162-06	5	8	1	4	2	4	-	-	-	-	-	-	-	24	19	6.229	0.328
CARDINAL HEALTH - CO	392-01	6	44	2	2	-	1	-	-	-	-	-	-	-	55	49	2.804	0.057
CARDINAL HEALTH - CO	392-03	6	38	2	2	-	1	-	-	-	-	-	-	-	49	43	2.688	0.063
CARDINAL HEALTH - CT	00163	6	29	8	-	-	-	-	-	-	-	-	-	-	43	37	1.839	0.050
CARDINAL HEALTH - CT	01567	-	2	1	-	-	-	-	-	-	-	-	-	-	3	3	0.218	0.073
CARDINAL HEALTH - FL	1264-9	3	14	4	5	-	-	-	-	-	-	-	-	-	26	23	2.937	0.128
CARDINAL HEALTH - FL	3273-7	2	4	1	-	-	-	-	-	-	-	-	-	-	7	5	0.228	0.046
CARDINAL HEALTH - FL	3453-1	6	23	9	-	-	-	-	-	-	-	-	-	-	38	32	1.702	0.053
CARDINAL HEALTH - FL	3453-2	11	16	4	1	-	-	-	-	-	-	-	-	-	32	21	1.451	0.069
CARDINAL HEALTH - FL	3453-3	7	18	2	-	-	-	-	-	-	-	-	-	-	27	20	0.925	0.046

Note: Columns "No Meas. Exposure" through ">12.0" fall under the heading "Number of Individuals with Whole Body Doses in the Ranges (rems)".

APPENDIX A
Table A - Annual TEDE for Agreement State Licensees
2008 (continued)

PROGRAM CODE - LICENSEE NAME	LICENSE#	No Meas. Exposure	Meas. <0.10	Number of Individuals with Whole Body Doses in the Ranges (rems)*											Total Number Monitored	Number with Meas. Dose	Total Collective TEDE (person-rem)	Average Meas. TEDE (rem)
				0.10-0.25	0.25-0.50	0.50-0.75	0.75-1.00	1.00-2.00	2.00-3.00	3.00-4.00	4.00-5.00	5.00-6.00	6.00-12.00	>12.0				
MANUFACTURING AND DISTRIBUTION – NUCLEAR PHARMACIES – 02500																		
CARDINAL HEALTH - FL	3453-5	7	13	4	2	1	1	-	-	-	-	-	-	-	28	21	3.168	0.151
CARDINAL HEALTH - FL	3453-6	8	21	5	-	-	-	-	-	-	-	-	-	-	34	26	1.136	0.044
CARDINAL HEALTH - FL	3453-7	5	19	7	5	-	-	-	-	-	-	-	-	-	36	31	3.465	0.112
CARDINAL HEALTH - FL	3453-3	10	22	9	-	1	-	-	-	-	-	-	-	-	42	32	3.023	0.094
CARDINAL HEALTH - FL	3453-9	1	19	4	2	1	-	-	-	-	-	-	-	-	26	25	1.704	0.068
CARDINAL HEALTH - GA	GA-4E7-1MD	3	15	3	4	1	-	-	-	-	-	-	-	-	26	23	2.863	0.124
CARDINAL HEALTH - GA	GA-4E7-2MD	7	6	-	-	-	-	-	-	-	-	-	-	-	13	6	0.120	0.020
CARDINAL HEALTH - GA	GA-4E7-3MD	1	4	-	-	-	-	-	-	-	-	-	-	-	5	4	0.019	0.005
CARDINAL HEALTH - GA	GA-823-2MD	2	9	2	1	-	-	-	-	-	-	-	-	-	14	12	0.733	0.061
CARDINAL HEALTH - IA	0043-1-77-NP	9	22	-	-	-	-	-	-	-	-	-	-	-	31	22	0.407	0.019
CARDINAL HEALTH - IL	IL-01721-01	11	49	5	4	-	-	-	-	-	-	-	-	-	69	58	3.651	0.063
CARDINAL HEALTH - IN	R0358-45	8	31	5	-	-	-	-	-	-	-	-	-	-	44	36	1.400	0.039
CARDINAL HEALTH - KS	20-C495-01	11	15	2	1	1	-	-	-	-	-	-	-	-	30	19	1.551	0.082
CARDINAL HEALTH - KY	202-204-32	8	35	3	-	1	-	-	-	-	-	-	-	-	46	38	1.040	0.027
CARDINAL HEALTH - KY	202-206-32	1	30	-	1	1	-	-	-	-	-	-	-	-	33	32	2.059	0.064
CARDINAL HEALTH - KY	202-333-32	1	3	-	-	-	-	-	-	-	-	-	-	-	4	3	0.041	0.014
CARDINAL HEALTH - LA	LA-10217-L01	6	1	-	-	-	-	-	-	-	-	-	-	-	7	1	0.034	0.034
CARDINAL HEALTH - LA	LA-10336-L01	4	6	-	-	-	-	-	-	-	-	-	-	-	10	6	0.176	0.029
CARDINAL HEALTH - LA	LA-3385-L01	7	13	2	-	-	-	-	-	-	-	-	-	-	22	15	0.608	0.041
CARDINAL HEALTH - LA	LA-5115-L01	1	12	10	2	-	-	-	-	-	-	-	-	-	25	24	2.740	0.114
CARDINAL HEALTH - LA	LA-5119-L01	4	8	-	-	-	-	-	-	-	-	-	-	-	12	8	0.139	0.017
CARDINAL HEALTH - LA	LA-5394-L01	3	15	4	-	1	-	-	-	-	-	-	-	-	23	20	1.574	0.079
CARDINAL HEALTH - LA	LA-7096-L01	4	6	-	-	-	-	-	-	-	-	-	-	-	10	6	0.118	0.020
CARDINAL HEALTH - MA	4008	-	1	-	1	1	3	-	1	-	-	-	-	-	7	7	8.647	1.235
CARDINAL HEALTH - MA	42-014E	8	29	11	12	4	4	1	-	-	-	-	-	-	69	61	13.811	0.226
CARDINAL HEALTH - MD	MD-05-48-01	4	13	-	2	-	-	-	-	-	-	-	-	-	19	15	1.048	0.070
CARDINAL HEALTH - MD	MD-33-77-01	2	1	3	1	1	-	-	-	-	-	-	-	-	8	6	1.546	0.258
CARDINAL HEALTH - MD	MD-33-198-01	16	19	11	5	3	-	3	-	-	-	-	-	-	57	41	9.591	0.234
CARDINAL HEALTH - MI	1549-4	10	19	4	3	1	-	-	-	-	-	-	-	-	37	27	2.578	0.095
CARDINAL HEALTH - MN	1037-205-89	24	40	3	6	-	-	-	-	-	-	-	-	-	73	49	2.778	0.057
CARDINAL HEALTH - MN	1045-100-89	-	-	1	2	1	3	-	-	-	-	-	-	-	7	7	3.931	0.562
CARDINAL HEALTH - MO	2840	17	37	5	2	-	-	-	-	-	-	-	-	-	61	44	2.138	0.049
CARDINAL HEALTH - MO	6141	-	1	1	2	-	-	-	-	-	-	-	-	-	4	4	0.928	0.232

APPENDIX A

Table A - Annual TEDE for Agreement State Licensees

2008 (continued)

MANUFACTURING AND DISTRIBUTION – NUCLEAR PHARMACIES – 02500

PROGRAM CODE - LICENSEE NAME	LICENSE#	No Meas. Exposure	Meas. <0.10	Number of Individuals with Whole Body Doses in the Ranges (rems)*											Total Number Monitored	Number with Meas. Dose	Total Collective TEDE (person-rem)	Average Meas. TEDE (rem)
				0.10-0.25	0.25-0.50	0.50-0.75	0.75-1.00	1.00-2.00	2.00-3.00	3.00-4.00	4.00-5.00	5.00-6.00	6.00-12.00	>12.0				
CARDINAL HEALTH - MS	MS-493-01	2	10	3	3	3	1	-	-	-	-	-	-	-	22	20	4.643	0.232
CARDINAL HEALTH - MS	MS-781-01	-	7	-	1	-	-	-	-	-	-	-	-	-	8	8	0.665	0.086
CARDINAL HEALTH - MS	MS-924-01	-	7	1	-	-	-	-	-	-	-	-	-	-	8	8	0.298	0.037
CARDINAL HEALTH - MS	MS-974-01	-	-	1	-	1	1	-	-	-	-	-	-	-	3	3	1.926	0.642
CARDINAL HEALTH - NC	011-0794-7	2	9	1	1	1	-	-	-	-	-	-	-	-	14	12	1.460	0.122
CARDINAL HEALTH - NC	025-0794-10	5	9	2	1	-	-	-	-	-	-	-	-	-	17	13	0.780	0.065
CARDINAL HEALTH - NC	026-0794-8	-	11	2	-	-	-	-	-	-	-	-	-	-	13	13	0.673	0.052
CARDINAL HEALTH - NC	041-0794-5	4	16	2	-	-	-	-	-	-	-	-	-	-	22	18	0.712	0.040
CARDINAL HEALTH - NC	060-0794-1	13	17	7	1	3	1	1	-	-	-	-	-	-	43	30	6.141	0.205
CARDINAL HEALTH - NE	01-65-01	9	30	5	-	-	-	-	-	-	-	-	-	-	44	35	1.413	0.040
CARDINAL HEALTH - NM	RP396-02	7	9	-	-	-	-	-	-	-	-	-	-	-	16	9	0.234	0.026
CARDINAL HEALTH - NV	03-11-0150-01	5	13	7	-	-	-	-	-	-	-	-	-	-	25	20	1.263	0.063
CARDINAL HEALTH - NV	092-0794-6	7	25	6	5	-	-	-	-	-	-	-	-	-	43	36	3.707	0.103
CARDINAL HEALTH - NY	C2328	9	6	3	2	1	-	-	-	-	-	-	-	-	21	12	1.817	0.151
CARDINAL HEALTH - NY	C2364	22	15	3	-	-	-	-	-	-	-	-	-	-	40	18	0.882	0.049
CARDINAL HEALTH - NY	C2449	2	12	4	1	-	-	-	-	-	-	-	-	-	19	17	1.347	0.079
CARDINAL HEALTH - NY	C2588	7	15	-	-	-	-	-	-	-	-	-	-	-	22	15	0.301	0.020
CARDINAL HEALTH - NY	C2593	4	23	1	-	-	-	-	-	-	-	-	-	-	28	24	0.648	0.027
CARDINAL HEALTH - NY	C2613	2	17	1	-	-	-	-	-	-	-	-	-	-	20	18	0.684	0.038
CARDINAL HEALTH - NY	C3046	9	29	3	-	-	-	-	-	-	-	-	-	-	41	32	1.060	0.033
CARDINAL HEALTH - NY	C3153	-	4	1	-	1	1	1	-	-	-	-	-	-	8	8	2.031	0.254
CARDINAL HEALTH - OH	02500180000	2	26	3	-	-	-	-	-	-	-	-	-	-	31	29	0.845	0.029
CARDINAL HEALTH - OH	02500250000	11	28	11	6	1	1	-	-	-	-	-	-	-	58	47	5.420	0.115
CARDINAL HEALTH - OH	02500310000	1	2	3	6	2	1	-	-	-	-	-	-	-	15	14	5.488	0.392
CARDINAL HEALTH - OH	02500490001	24	14	5	1	-	-	-	-	-	-	-	-	-	44	20	1.552	0.078
CARDINAL HEALTH - OH	02500580000	6	21	4	1	-	-	-	-	-	-	-	-	-	32	26	1.163	0.045
CARDINAL HEALTH - OH	02500790000	3	10	-	-	-	-	-	-	-	-	-	-	-	13	10	0.167	0.017
CARDINAL HEALTH - OH	03214250004	-	-	-	2	1	1	-	-	-	-	-	-	-	4	4	1.903	0.476
CARDINAL HEALTH - OH	03214310000	-	-	1	1	1	1	1	-	-	-	-	-	-	5	5	3.570	0.714
CARDINAL HEALTH - OH	4853-19	7	15	2	1	-	-	1	-	-	-	-	-	-	26	19	2.473	0.130
CARDINAL HEALTH - OK	OK-19583-02MD	4	22	3	1	-	-	-	-	-	-	-	-	-	30	26	1.138	0.044
CARDINAL HEALTH - OK	OK-23359-02MD	7	15	-	-	-	-	-	-	-	-	-	-	-	22	15	0.250	0.017
CARDINAL HEALTH - OR	ORE-90509	1	11	2	3	-	-	-	-	-	-	-	-	-	17	16	1.779	0.111

APPENDIX A

Table A - Annual TEDE for Agreement State Licensees
2008 (continued)

MANUFACTURING AND DISTRIBUTION – NUCLEAR PHARMACIES – 02500

PROGRAM CODE - LICENSEE NAME	LICENSE#	No Meas. Exposure	Meas. <0.10	0.10-0.25	0.25-0.50	0.50-0.75	0.75-1.00	1.00-2.00	2.00-3.00	3.00-4.00	4.00-5.00	5.00-6.00	6.00-12.00	>12.0	Total Number Monitored	Number with Meas. Dose	Total Collective TEDE (person-rem)	Average Meas. TEDE (rem)
CARDINAL HEALTH - OR	ORE-90703	6	16	2	2	-	-	-	-	-	-	-	-	-	26	20	1.558	0.078
CARDINAL HEALTH - OR	ORE-90914	11	18	6	4	2	1	-	-	-	-	-	-	-	42	31	5.039	0.163
CARDINAL HEALTH - PA	10239-02-002	2	16	4	3	4	-	1	-	-	-	-	-	-	30	28	5.946	0.212
CARDINAL HEALTH - PA	AC5C-58988	-	2	3	-	-	-	-	-	-	-	-	-	-	5	5	0.489	0.098
CARDINAL HEALTH - PA	PA-0384	6	48	11	1	3	-	-	-	-	-	-	-	-	69	63	4.916	0.078
CARDINAL HEALTH - PA	PA-0385	4	43	9	5	5	-	-	-	-	-	-	-	-	66	62	7.046	0.114
CARDINAL HEALTH - PA	PA-0415	3	25	3	-	-	-	-	-	-	-	-	-	-	31	28	1.239	0.044
CARDINAL HEALTH - PA	PA-0460	3	12	3	-	-	-	-	-	-	-	-	-	-	18	15	0.591	0.039
CARDINAL HEALTH - PA	PA-0643	4	14	-	1	-	-	-	-	-	-	-	-	-	19	15	0.802	0.053
CARDINAL HEALTH - PA	PA-0680	6	21	4	-	-	-	-	-	-	-	-	-	-	31	25	1.046	0.042
CARDINAL HEALTH - RI	3B-114-01	8	34	2	-	-	-	-	-	-	-	-	-	-	44	36	1.112	0.031
CARDINAL HEALTH - SC	448	4	19	3	1	-	-	-	-	-	-	-	-	-	27	23	1.566	0.068
CARDINAL HEALTH - SD	47111	-	13	2	-	-	-	-	-	-	-	-	-	-	15	15	1.065	0.071
CARDINAL HEALTH - TN	R-19199-B15	9	20	2	-	-	-	-	-	-	-	-	-	-	31	22	0.784	0.036
CARDINAL HEALTH - TN	R-33111-I14	1	21	1	-	-	-	-	-	-	-	-	-	-	23	22	0.414	0.019
CARDINAL HEALTH - TN	R47080-L14	12	17	5	-	-	-	-	-	-	-	-	-	-	34	22	1.079	0.049
CARDINAL HEALTH - TN	R-57025-C15	4	3	1	-	-	-	-	-	-	-	-	-	-	8	4	0.236	0.059
CARDINAL HEALTH - TN	R-71029-B11	2	6	-	-	-	-	-	-	-	-	-	-	-	8	6	0.122	0.020
CARDINAL HEALTH - TN	R-79174-D17	3	12	5	2	-	-	-	-	-	-	-	-	-	22	19	2.094	0.110
CARDINAL HEALTH - TX	L0-1911	14	39	12	7	1	-	-	-	-	-	-	-	-	73	59	5.536	0.094
CARDINAL HEALTH - TX	L0-1993	1	6	-	-	-	-	-	-	-	-	-	-	-	7	6	0.095	0.016
CARDINAL HEALTH - TX	L0-2033	2	34	10	3	1	1	-	-	-	-	-	-	-	51	49	4.757	0.097
CARDINAL HEALTH - TX	L0-2048	5	29	4	3	1	-	-	-	-	-	-	-	-	42	37	3.340	0.090
CARDINAL HEALTH - TX	L0-2117	2	20	1	1	-	-	-	-	-	-	-	-	-	24	22	0.715	0.033
CARDINAL HEALTH - TX	L0-2737	3	3	-	-	-	-	-	-	-	-	-	-	-	6	3	0.063	0.021
CARDINAL HEALTH - TX	L0-3396	4	5	1	-	-	-	-	-	-	-	-	-	-	10	6	0.230	0.038
CARDINAL HEALTH - TX	L0-4043	1	5	1	-	-	-	-	-	-	-	-	-	-	7	6	0.179	0.030
CARDINAL HEALTH - TX	L0-4573	22	29	6	4	-	-	-	-	-	-	-	-	-	61	39	3.120	0.080
CARDINAL HEALTH - TX	L0-4781	10	15	1	-	-	-	-	-	-	-	-	-	-	26	16	0.387	0.024
CARDINAL HEALTH - TX	L0-4785	5	14	1	-	-	-	-	-	-	-	-	-	-	20	15	0.398	0.027
CARDINAL HEALTH - TX	L0-5461	-	7	-	-	-	-	-	-	-	-	-	-	-	7	7	0.192	0.027
CARDINAL HEALTH - TX	L0-5536	-	-	1	2	-	-	-	-	-	-	-	-	-	3	3	0.804	0.268
CARDINAL HEALTH - TX	L0-5610	1	-	-	2	1	-	1	-	-	-	-	-	-	5	4	2.350	0.588

APPENDIX A
Table A - Annual TEDE for Agreement State Licensees
2008 (continued)

PROGRAM CODE - LICENSEE NAME	LICENSE#	No Meas. Exposure	Meas. <0.10	Number of Individuals with Whole Body Doses in the Ranges (rems)*											Total Number Monitored	Number with Meas. Dose	Total Collective TEDE (person-rem)	Average Meas. TEDE (rem)
				0.10-0.25	0.25-0.50	0.50-0.75	0.75-1.00	1.00-2.00	2.00-3.00	3.00-4.00	4.00-5.00	5.00-6.00	6.00-12.00	>12.0				
MANUFACTURING AND DISTRIBUTION – NUCLEAR PHARMACIES – 02500																		
CARDINAL HEALTH - UNITED STATES	R-79272-111	-	2	2	1	-	-	-	-	-	-	-	-	-	5	5	0.718	0.144
CARDINAL HEALTH - US	065-0794-11	-	7	3	-	-	-	-	-	-	-	-	-	-	10	10	0.689	0.069
CARDINAL HEALTH - US	CH-US-001	22	33	9	9	3	2	1	-	-	-	-	-	-	79	57	10.172	0.178
CARDINAL HEALTH - UT	UT-1800225	4	20	3	-	-	-	-	-	-	-	-	-	-	27	23	1.287	0.056
CARDINAL HEALTH - VA	760-34-1	13	41	9	1	-	-	-	-	-	-	-	-	-	64	51	2.515	0.049
CARDINAL HEALTH - WA	WN-NP003-1	2	27	5	2	-	-	-	-	-	-	-	-	-	36	34	1.983	0.058
CARDINAL HEALTH - WA	WN-NP004-1	1	19	4	1	-	-	-	-	-	-	-	-	-	25	24	1.458	0.061
CARDINAL HEALTH - WA	WN-NP005-1	5	15	1	-	-	-	-	-	-	-	-	-	-	21	16	0.551	0.034
CARDINAL HEALTH - WA	WN-NP006-1	-	4	1	1	-	-	-	-	-	-	-	-	-	6	6	0.528	0.088
CARDINAL HEALTH - WA	WN-NP011-1	49	24	9	15	2	-	1	-	-	-	-	-	-	100	51	10.119	0.198
CARDINAL HEALTH - WI	025-1123-01	2	13	4	1	-	-	-	-	-	-	-	-	-	20	18	1.333	0.074
CARDINAL HEALTH - WI	087-1312-01	8	17	5	-	-	-	-	-	-	-	-	-	-	30	22	0.978	0.044
CARDINAL HEALTH - WI	141-1306-01	3	10	1	-	-	-	-	-	-	-	-	-	-	14	11	0.381	0.035
GE HEALTHCARE - ANAHEIM	4810-30	20	10	1	-	-	-	-	-	-	-	-	-	-	31	11	0.514	0.047
GE HEALTHCARE - SACRAMENTO	4809-34	21	8	-	-	-	-	-	-	-	-	-	-	-	29	8	0.415	0.052
GE HEALTHCARE - SAN DIEGO	5796-37	17	6	-	-	-	-	-	-	-	-	-	-	-	23	6	0.200	0.033
GE HEALTHCARE - SAN JOSE	4811-43	8	6	4	3	-	-	-	-	-	-	-	-	-	21	13	2.006	0.154
GE HEALTHCARE - VAN NUYS	5143-19	6	17	-	-	-	-	-	-	-	-	-	-	-	23	17	0.563	0.033
IBA MOLECULAR NORTH AMERICA, INC.	CA-7131-43	9	21	3	2	4	-	3	1	1	-	-	-	-	44	35	14.567	0.416
MALLINCKRODT, INC.	859-01	9	4	1	-	-	-	-	-	-	-	-	-	-	14	5	0.277	0.055
MORAVEK BIOCHEMICALS, INC.	2960-30	3	28	6	3	-	-	-	-	-	-	-	-	-	41	38	4.800	0.126
WYLE LABORATORIES	FL-2953-1	20	9	-	-	-	-	-	-	-	-	-	-	-	29	9	0.203	0.023
Total	154	893	2,451	477	219	71	28	21	3	1	-	-	-	-	4,164	3,271	318.430	0.097
WELL LOGGING BYPRODUCT AND/OR SNM SEALED SOURCES ONLY - 03111																		
ISOTOPE PRODUCTS LABS	1509-19	5	20	13	7	6	4	14	1	-	-	-	-	-	70	65	34.384	0.529
Total	1	5	20	13	7	6	4	14	1	-	-	-	-	-	70	65	34.384	0.529
MEASURING SYSTEMS PORTABLE GAUGES - 03121																		
CPN INTERNATIONAL, INC.	1100-07	-	6	4	2	-	-	-	-	-	-	-	-	-	12	12	1.595	0.133
Total	1	-	6	4	2	-	-	-	-	-	-	-	-	-	12	12	1.595	0.133

APPENDIX A

Table A - Annual TEDE for Agreement State Licensees
2008 (continued)

PROGRAM CODE - LICENSEE NAME	LICENSE#	No Meas. Exposure	Meas. <0.10	0.10-0.25	0.25-0.50	0.50-0.75	0.75-1.00	1.00-2.00	2.00-3.00	3.00-4.00	4.00-5.00	5.00-6.00	6.00-12.00	>12.0	Total Number Monitored	Number with Meas. Dose	Total Collective TEDE (person-rem)	Average Meas. TEDE (rem)
MANUFACTURING AND DISTRIBUTION – OTHER - 03214																		
HOCHIKI AMERICA CORPORATION	2090-30	20	-	-	-	-	-	-	-	-	-	-	-	-	20	-	-	-
J L. SHEPHERD AND ASSOCIATES	CA 1777-19	15	6	1	1	-	-	-	-	-	-	-	-	-	23	8	0.770	0.096
Total	2	35	6	1	1	-	-	-	-	-	-	-	-	-	43	8	0.770	0.096
OTHER SERVICES - 03225																		
HYSPAN PRECISION PRODUCTS, INC.	7010-37	-	3	-	-	-	-	-	-	-	-	-	-	-	3	3	-	-
Total	1	-	3	-	-	-	-	-	-	-	-	-	-	-	3	3	-	-
WASTE DISPOSAL SERVICE PROCESSING AND/OR REPACKAGING - 03234																		
ENVIRONMENTAL MANAGEMENT & CONTROLS	3546-50	-	-	2	2	1	-	-	-	-	-	-	-	-	5	5	1.717	0.343
Total	1	-	-	2	2	1	-	-	-	-	-	-	-	-	5	5	1.717	0.343
INDUSTRIAL RADIOGRAPHY – FIXED LOCATION - 03310																		
LAFAYETTE TESTING SERVICES, INC.	079-1147-01	1	8	1	-	-	-	-	-	-	-	-	-	-	10	9	0.498	0.055
METAL TEK - WISCONSIN CENTRIFUGAL DIVISION	133-1131-01	4	1	2	2	-	-	-	-	-	-	-	-	-	9	5	1.372	0.274
SONGS - RADIOGRAPHY GROUP	CA-5244-30	11	-	-	-	-	-	-	-	-	-	-	-	-	11	-	-	-
WAUKESHA FOUNDRY, INC - WI	133-1337-01	4	-	-	-	-	-	-	-	-	-	-	-	-	4	-	-	-
Total	4	20	9	3	2	-	-	-	-	-	-	-	-	-	34	14	1.870	0.134
INDUSTRIAL RADIOGRAPHY – TEMPORARY JOB SITE – 03320																		
ACUREN	133-2008-01	-	18	6	19	6	9	19	4	-	-	-	-	-	81	81	55.109	0.680
ADVEX CORPORATION	650-25-1	2	5	1	1	1	-	-	-	-	-	-	-	-	10	8	1.337	0.167
ALARON CORPORATION	PA-0678	13	64	18	2	1	-	6	4	-	-	-	-	-	108	95	23.507	0.247
ALPHA-OMEGA SERVICES, INC.	3925-1C	3	16	2	1	-	-	-	-	-	-	-	-	-	22	19	0.769	0.040
AMERICAN ENGINEERING TESTING INC	073-2012-02	1	1	1	1	-	1	1	3	1	-	-	-	-	10	9	12.518	1.391
BRAUN INTERTEC - MN	1082-1C0-27	4	7	3	6	3	5	3	2	-	-	-	-	-	33	29	18.129	0.625
CONAM INSPECTION - UTAH	UT-1800485	-	1	-	-	-	-	-	-	-	-	-	-	-	1	1	0.010	0.010
CONSTRUCTION MATERIALS TESTING, INC.	0799-07	2	3	1	2	-	-	1	-	-	-	-	-	-	9	7	1.962	0.280
COOPERHEAT-MQS INSPECTION	388-01	11	9	6	8	3	4	7	1	-	-	-	-	-	49	38	21.206	0.558
EDGE INSPECTION GROUP, INC.	CA-7214-48	-	6	3	-	4	2	5	1	-	-	-	-	-	21	21	15.532	0.740
GREAT LAKES TESTING, INC.	009-1115-01	2	6	2	4	4	3	-	-	-	-	-	-	-	21	19	7.199	0.379

APPENDIX A
Table A - Annual TEDE for Agreement State Licensees
2008 (continued)

PROGRAM CODE - LICENSEE NAME	LICENSE#	No. Meas. Exposure	Meas. <0.10	0.10-0.25	0.25-0.50	0.50-0.75	0.75-1.00	1.00-2.00	2.00-3.00	3.00-4.00	4.00-5.00	5.00-6.00	6.00-12.00	>12.0	Total Number Monitored	Number with Meas. Dose	Total Collective TEDE (person-rem)	Average Meas. TEDE (rem)
INDUSTRIAL RADIOGRAPHY – TEMPORARY JOB SITE – 03320																		
INTERNATIONAL INSPECTION, INC.	5046-19	3	6	4	-	-	-	-	-	-	-	-	-	1		14	11	9.323	0.848
MIDWEST INDUSTRIAL X-RAY, INC.	ND-33-27427	-	1	1	1	2	2	8	1	-	1	-	-		17	17	21.759	1.280
NDT LABORATORIES, INC.	1651-43	-	3	1	1	-	-	-	-	-	-	-	-		5	5	0.523	0.105
NDT SPECIALISTS, INC.	079-1199-01	-	2	4	-	-	2	2	-	1	-	-	-		11	11	8.304	0.755
SUBSURFACE IMAGING, INC.	7449-19	2	5	10	-	-	-	-	-	-	-	-	-		17	15	2.071	0.138
TC INSPECTION, INC.	5299-07	30	14	20	12	7	8	6	1	-	-	-	-		98	68	31.067	0.457
TEAM INDUSTRIAL SERVICES, INC	079-2005-01	2	11	2	2	1	1	1	1	-	-	-	-		21	19	6.999	0.368
TECH CORR INSPECTION & ENGINEERING	TX-L05972	1	-	-	-	-	-	-	-	-	-	-	-		1	-	-	-
TESTING ENGINEERS, INC.	1898-01	-	2	1	1	1	-	-	-	-	-	-	-		5	5	1.186	0.237
THREE RIVERS GAMMA SERVICES	PA-1165	-	-	-	-	2	-	-	-	-	-	-	-		2	2	1.962	0.981
TWIN PORTS TESTING, INC.	031-1317-02	6	2	2	3	1	1	1	-	-	-	-	-		16	10	4.979	0.498
VALLEY INDUSTRIAL X-RAY & INSPECTION	CA-4182-15	9	40	14	15	15	16	29	7	1	1	-	-		146	137	91.811	0.670
WELDSONIX, INC.	L05718	7	15	23	21	17	7	5	1	1	-	-	-		97	90	41.157	0.457
WESTERN INDUSTRIAL X-RAY	4424-48	-	-	-	2	-	1	3	-	-	-	-	-		8	8	5.138	0.642
WESTERN X-RAY CORPORATION	5324-56	4	1	1	1	1	-	-	-	-	-	-	-		9	5	2.168	0.434
YUBA HEAT TRANSFER	OK-13735-01	-	1	3	-	-	-	-	-	-	-	-	-		4	4	0.347	0.087
Total	**27**	102	239	131	102	69	65	98	24	4	1	-	1	-	836	734	386.072	0.526
IRRADIATORS - OTHER - LESS THAN 10000 CURIES - 03511																		
INDUSTRIAL NUCLEAR CO., INC.	2229-01	6	-	1	1	-	1	-	-	-	-	-	-		10	4	2.575	0.644
Total	**1**	6	-	1	1	-	1	-	-	-	-	-	-	-	10	4	2.575	0.644

APPENDIX A
Table A - Annual TEDE for Agreement State Licensees
2009

MANUFACTURING AND DISTRIBUTION – NUCLEAR PHARMACIES – 02500

PROGRAM CODE - LICENSEE NAME	LICENSE#	No. Meas. Exposure	Meas. <0.10	0.10-0.25	0.25-0.50	0.50-0.75	0.75-1.00	1.00-2.00	2.00-3.00	3.00-4.00	4.00-5.00	5.00-6.00	6.00-12.00	>12.0	Total Number Monitored	Number with Meas. Dose	Total Collective TEDE (person-rem)	Average Meas. TEDE (rem)
CARDINAL HEALTH - AK	0707RA90T-912	1	3	1	-	-	-	-	-	-	-	-	-	-	5	4	0.201	0.050
CARDINAL HEALTH - AL	1068	-	4	1	-	-	-	-	-	-	-	-	-	-	5	5	0.187	0.037
CARDINAL HEALTH - AL	1168	2	37	8	2	1	-	-	-	-	-	-	-	-	50	48	3.504	0.073
CARDINAL HEALTH - AL	1290	1	1	-	-	-	-	-	-	-	-	-	-	-	2	1	0.051	0.051
CARDINAL HEALTH - AR	ARK-642-02500	28	45	3	-	-	-	-	-	-	-	-	-	-	76	48	1.831	0.038
CARDINAL HEALTH - AZ	01-084	-	7	-	2	-	-	-	-	-	-	-	-	-	9	9	0.872	0.097
CARDINAL HEALTH - AZ	07-123	10	35	14	5	-	-	-	-	-	-	-	-	-	64	54	4.549	0.084
CARDINAL HEALTH - AZ	07-529	1	1	3	2	-	1	2	2	-	-	-	-	-	12	11	10.085	0.917
CARDINAL HEALTH - AZ	08-036	-	3	-	1	-	-	-	-	-	-	-	-	-	4	4	0.545	0.136
CARDINAL HEALTH - CA	2891-37	-	14	7	1	-	-	-	-	-	-	-	-	-	22	22	1.924	0.087
CARDINAL HEALTH - CA	3317-19	3	15	-	-	-	-	-	-	-	-	-	-	-	18	15	0.219	0.015
CARDINAL HEALTH - CA	3426-43	5	8	1	-	-	-	-	-	-	-	-	-	-	14	9	0.235	0.026
CARDINAL HEALTH - CA	3469-1	-	5	2	-	-	-	-	-	-	-	-	-	-	7	7	0.608	0.087
CARDINAL HEALTH - CA	3469-2	-	-	-	-	1	2	1	1	-	-	-	-	-	5	5	6.460	1.292
CARDINAL HEALTH - CA	3673-34	3	34	4	-	-	-	-	-	-	-	-	-	-	41	38	1.164	0.031
CARDINAL HEALTH - CA	3822-19	7	37	15	3	-	-	-	-	-	-	-	-	-	62	55	5.276	0.096
CARDINAL HEALTH - CA	3832-01	6	26	3	-	-	-	-	-	-	-	-	-	-	35	29	1.318	0.045
CARDINAL HEALTH - CA	4905-10	6	26	2	-	-	-	-	-	-	-	-	-	-	34	28	0.841	0.030
CARDINAL HEALTH - CA	4999-30	4	27	6	1	-	-	-	-	-	-	-	-	-	38	34	1.610	0.047
CARDINAL HEALTH - CA	5218-36	4	16	1	2	3	1	-	-	-	-	-	-	-	27	23	3.980	0.173
CARDINAL HEALTH - CA	5905-15	1	8	-	-	-	-	-	-	-	-	-	-	-	9	8	0.219	0.027
CARDINAL HEALTH - CA	5910-50	1	5	1	-	-	-	-	-	-	-	-	-	-	7	6	0.187	0.031
CARDINAL HEALTH - CA	6321-45	4	8	1	1	-	-	-	-	-	-	-	-	-	14	10	0.668	0.067
CARDINAL HEALTH - CA	6691-33	3	7	-	-	-	-	-	-	-	-	-	-	-	10	7	0.155	0.022
CARDINAL HEALTH - CA	6924-36	-	1	-	3	-	1	2	-	-	-	-	-	-	7	7	4.932	0.705
CARDINAL HEALTH - CA	6925-19	-	5	2	5	1	-	-	-	-	-	-	-	-	13	13	2.704	0.208
CARDINAL HEALTH - CO	162-06	4	14	2	5	1	3	-	-	-	-	-	-	-	29	25	5.860	0.234
CARDINAL HEALTH - CO	392-01	8	41	7	2	1	-	-	-	-	-	-	-	-	59	51	4.305	0.084
CARDINAL HEALTH - CO	392-03	7	36	7	2	-	-	-	-	-	-	-	-	-	52	45	2.898	0.064
CARDINAL HEALTH - CT	00163	4	35	9	-	-	-	-	-	-	-	-	-	-	48	44	2.431	0.055
CARDINAL HEALTH - CT	015567	-	-	-	1	1	2	1	-	-	-	-	-	-	5	5	4.024	0.805

APPENDIX A

Table A - Annual TEDE for Agreement State Licensees
2009 (continued)[*]

MANUFACTURING AND DISTRIBUTION – NUCLEAR PHARMACIES – 02500

PROGRAM CODE - LICENSEE NAME	LICENSE#	No Meas. Exposure	Number of Individuals with Whole Body Doses in the Ranges (rems)[*]												Total Number Monitored	Number with Meas. Dose	Total Collective TEDE (person-rem)	Average Meas. TEDE (rem)
			Meas. <0.10	0.10-0.25	0.25-0.50	0.50-0.75	0.75-1.00	1.00-2.00	2.00-3.00	3.00-4.00	4.00-5.00	5.00-6.00	6.00-12.00	>12.0				
CARDINAL HEALTH - FL	1264-9	1	17	6	2	-	-	-	-	-	-	-	-	-	26	25	2.214	0.089
CARDINAL HEALTH - FL	3273-7	1	4	-	-	-	-	-	-	-	-	-	-	-	5	4	0.037	0.009
CARDINAL HEALTH - FL	3453-1	8	28	7	1	-	-	-	-	-	-	-	-	-	44	36	1.639	0.046
CARDINAL HEALTH - FL	3453-13	-	2	1	-	-	2	-	-	-	-	-	-	-	5	5	2.051	0.410
CARDINAL HEALTH - FL	3453-2	7	21	2	-	-	-	-	-	-	-	-	-	-	30	23	0.808	0.035
CARDINAL HEALTH - FL	3453-3	10	19	2	-	-	-	-	-	-	-	-	-	-	31	21	0.544	0.026
CARDINAL HEALTH - FL	3453-5	7	8	6	3	-	2	-	-	-	-	-	-	-	26	19	4.184	0.220
CARDINAL HEALTH - FL	3453-6	19	20	2	-	-	-	-	-	-	-	-	-	-	41	22	0.821	0.037
CARDINAL HEALTH - FL	3453-7	4	28	10	2	-	-	-	-	-	-	-	-	-	44	40	3.297	0.082
CARDINAL HEALTH - FL	3453-8	3	28	6	1	1	1	-	-	-	-	-	-	-	40	37	3.541	0.096
CARDINAL HEALTH - FL	3453-9	4	18	3	1	-	-	-	1	-	-	-	-	-	27	23	4.193	0.182
CARDINAL HEALTH - GA	GA-467-1MD	4	17	5	-	-	-	-	-	-	-	-	-	-	27	23	1.646	0.072
CARDINAL HEALTH - GA	GA-467-2MD	3	8	-	-	-	-	-	-	-	-	-	-	-	11	8	0.062	0.008
CARDINAL HEALTH - GA	GA-467-3MD	2	3	-	-	-	-	-	-	-	-	-	-	-	5	3	0.042	0.014
CARDINAL HEALTH - GA	GA-823-2MD	2	11	2	-	-	-	-	-	-	-	-	-	-	15	13	0.644	0.050
CARDINAL HEALTH - IA	0043-1-77-NP	9	19	1	-	-	-	-	-	-	-	-	-	-	29	20	0.624	0.031
CARDINAL HEALTH - IL	IL-01721-01	9	47	5	3	-	-	-	-	-	-	-	-	-	64	55	3.198	0.058
CARDINAL HEALTH - IN	R0358-45	18	29	2	-	-	-	-	-	-	-	-	-	-	49	31	0.933	0.030
CARDINAL HEALTH - KS	20-C495-01	13	24	1	-	-	-	-	-	-	-	-	-	-	37	24	0.414	0.017
CARDINAL HEALTH - KY	202-204-32	8	29	1	-	-	-	-	-	-	-	-	-	-	38	30	0.519	0.017
CARDINAL HEALTH - KY	202-206-32	5	20	2	1	1	2	1	-	-	-	-	-	-	32	27	4.718	0.175
CARDINAL HEALTH - KY	202-333-32	-	4	-	-	-	-	-	-	-	-	-	-	-	4	4	0.091	0.023
CARDINAL HEALTH - LA	LA-10217-L01	2	5	1	-	-	-	-	-	-	-	-	-	-	8	6	0.248	0.041
CARDINAL HEALTH - LA	LA-10336-L01	-	7	-	-	-	-	-	-	-	-	-	-	-	7	7	0.212	0.030
CARDINAL HEALTH - LA	LA-3385-L01	10	12	1	-	-	-	-	-	-	-	-	-	-	23	13	0.445	0.034
CARDINAL HEALTH - LA	LA-5115-L01	-	13	4	1	-	-	-	-	-	-	-	-	-	18	18	1.206	0.067
CARDINAL HEALTH - LA	LA-5119-L01	5	6	-	-	-	-	-	-	-	-	-	-	-	11	6	0.092	0.015
CARDINAL HEALTH - LA	LA-5394-L01	1	14	5	2	-	-	-	-	-	-	-	-	-	22	21	2.011	0.096
CARDINAL HEALTH - LA	LA-7096-L01	4	5	-	-	-	-	-	-	-	-	-	-	-	9	5	0.134	0.027
CARDINAL HEALTH - MA	4008	-	5	1	-	2	-	2	2	-	-	-	-	-	12	12	8.611	0.718
CARDINAL HEALTH - MA	42-0146	3	25	19	9	4	2	-	-	-	-	-	-	-	62	59	11.329	0.192
CARDINAL HEALTH - MD	MD-05-148-01	8	-	-	-	-	-	-	-	-	-	-	-	-	8	-	-	-
CARDINAL HEALTH - MD	MD-33-177-01	-	3	4	1	1	-	-	-	-	-	-	-	-	9	9	1.833	0.204

APPENDIX A
Table A - Annual TEDE for Agreement State Licensees
2009 (continued)

MANUFACTURING AND DISTRIBUTION – NUCLEAR PHARMACIES – 02500

PROGRAM CODE - LICENSEE NAME	LICENSE#	No Meas. Exposure	Number of Individuals with Whole Body Doses in the Ranges (rems)*												Total Number Monitored	Number with Meas. Dose	Total Collective TEDE (person-rem)	Average Meas. TEDE (rem)
			Meas. <0.10	0.10-0.25	0.25-0.50	0.50-0.75	0.75-1.00	1.00-2.00	2.00-3.00	3.00-4.00	4.00-5.00	5.00-6.00	6.00-12.00	>12.0				
CARDINAL HEALTH - MD	MD-33-19E-01	7	39	18	7	5	-	1	-	-	-	-	-	-	77	70	11.724	0.167
CARDINAL HEALTH - ME	19233	1	3	1	-	-	-	-	-	-	-	-	-	-	5	4	0.141	0.035
CARDINAL HEALTH - MN	1037-205-E9	13	38	6	3	1	-	-	-	-	-	-	-	-	61	48	3.061	0.064
CARDINAL HEALTH - MN	1045-100-C9	-	-	-	2	2	2	-	-	-	-	-	-	-	6	6	3.865	0.644
CARDINAL HEALTH - MO	2840	23	48	6	3	2	-	-	-	-	-	-	-	-	80	57	2.705	0.047
CARDINAL HEALTH - MO	6141	-	-	2	-	-	-	-	-	-	-	-	-	-	4	4	1.087	0.272
CARDINAL HEALTH - MS	MS-493-01	2	12	4	3	2	-	-	-	-	-	-	-	-	23	21	3.242	0.154
CARDINAL HEALTH - MS	MS-781-01	-	3	1	1	-	-	-	-	-	-	-	-	-	5	5	0.417	0.083
CARDINAL HEALTH - MS	MS-924-01	1	4	-	-	-	-	-	-	-	-	-	-	-	5	4	0.128	0.032
CARDINAL HEALTH - MS	MS-974-01	-	1	1	1	1	-	1	-	-	-	-	-	-	4	4	1.967	0.492
CARDINAL HEALTH - NC	011-0794-7	3	5	2	1	1	-	-	-	-	-	-	-	-	12	9	1.347	0.150
CARDINAL HEALTH - NC	025-0794-10	2	13	1	1	-	-	-	-	-	-	-	-	-	17	15	0.653	0.044
CARDINAL HEALTH - NC	026-0794-8	-	12	3	-	-	-	-	-	-	-	-	-	-	15	15	0.662	0.044
CARDINAL HEALTH - NC	041-0794-5	5	10	2	-	-	-	-	-	-	-	-	-	-	17	12	0.581	0.048
CARDINAL HEALTH - NC	060-0794-1	4	25	5	4	2	-	3	-	-	-	-	-	-	43	39	7.986	0.205
CARDINAL HEALTH - NE	01-65-01	7	26	3	4	2	-	2	-	-	-	-	-	-	44	37	6.551	0.177
CARDINAL HEALTH - NM	AP-403-13	-	1	1	-	1	-	1	1	-	-	-	-	-	5	5	2.647	0.529
CARDINAL HEALTH - NM	RP396-02	5	16	3	1	-	-	-	-	-	-	-	-	-	25	20	1.262	0.063
CARDINAL HEALTH - NV	03-11-0150-01	4	13	3	-	-	-	-	-	-	-	-	-	-	20	16	0.847	0.053
CARDINAL HEALTH - NV	03-11-0332-01	1	1	12	3	-	3	-	-	-	-	-	-	-	21	20	6.151	0.308
CARDINAL HEALTH - NV	03-11-0505-01	-	-	-	2	1	-	-	-	-	-	-	-	-	3	3	1.443	0.481
CARDINAL HEALTH - NV	092-0794-6	2	26	3	3	-	-	-	-	-	-	-	-	-	34	32	2.303	0.072
CARDINAL HEALTH - NY	C2328	7	6	4	1	-	-	-	-	-	-	-	-	-	18	11	1.215	0.110
CARDINAL HEALTH - NY	C2364	12	20	3	1	-	-	-	-	-	-	-	-	-	36	24	1.066	0.044
CARDINAL HEALTH - NY	C2449	7	16	3	2	-	-	-	-	-	-	-	-	-	28	21	1.557	0.074
CARDINAL HEALTH - NY	C2588	5	17	-	-	-	-	-	-	-	-	-	-	-	22	17	0.393	0.023
CARDINAL HEALTH - NY	C2593	8	23	1	-	-	-	-	-	-	-	-	-	-	32	24	0.769	0.032
CARDINAL HEALTH - NY	C2613	4	14	1	-	-	-	-	-	-	-	-	-	-	19	15	0.649	0.043
CARDINAL HEALTH - NY	C3046	3	29	3	-	1	-	-	-	-	-	-	-	-	36	33	1.357	0.041
CARDINAL HEALTH - NY	C3153	1	4	-	3	1	-	-	-	-	-	-	-	-	9	8	1.812	0.227
CARDINAL HEALTH - OH	02500180000	5	20	2	1	-	-	-	-	-	-	-	-	-	28	23	1.013	0.044
CARDINAL HEALTH - OH	02500250000	14	30	9	3	2	1	-	-	-	-	-	-	-	59	45	5.058	0.112
CARDINAL HEALTH - OH	02500310000	2	8	2	1	-	-	-	-	-	-	-	-	-	13	11	1.259	0.114

APPENDIX A

Table A - Annual TEDE for Agreement State Licensees
2009 (continued)

MANUFACTURING AND DISTRIBUTION – NUCLEAR PHARMACIES – 02500

PROGRAM CODE - LICENSEE NAME	LICENSE#	No Meas. Exposure	Meas. <0.10	Number of Individuals with Whole Body Doses in the Ranges (rems)*											Total Number Monitored	Number with Meas. Dose	Total Collective TEDE (person-rem)	Average Meas. TEDE (rem)
				0.10-0.25	0.25-0.50	0.50-0.75	0.75-1.00	1.00-2.00	2.00-3.00	3.00-4.00	4.00-5.00	5.00-6.00	6.00-12.00	>12.0				
CARDINAL HEALTH - OH	02500490001	17	18	5	-	-	-	-	-	-	-	-	-	-	40	23	1.180	0.051
CARDINAL HEALTH - OH	02500580000	10	19	3	-	-	-	-	-	-	-	-	-	-	32	22	0.823	0.037
CARDINAL HEALTH - OH	02500790000	-	12	-	-	-	-	-	-	-	-	-	-	-	12	12	0.268	0.022
CARDINAL HEALTH - OH	03214250004	-	4	1	1	2	-	1	-	-	-	-	-	-	9	9	2.860	0.318
CARDINAL HEALTH - OH	03214310000	-	2	-	2	-	-	2	-	-	-	-	-	-	6	6	4.090	0.682
CARDINAL HEALTH - OH	4853-19	10	9	-	-	-	-	-	-	-	-	-	-	-	19	9	0.051	0.006
CARDINAL HEALTH - OK	OK-19583-02MD	14	16	-	-	-	-	-	-	-	-	-	-	-	30	16	0.331	0.021
CARDINAL HEALTH - OK	OK-23359-02MD	4	21	-	-	-	-	-	-	-	-	-	-	-	25	21	0.360	0.017
CARDINAL HEALTH - OR	ORE-90509	1	9	4	-	-	2	-	-	-	-	-	-	-	20	19	4.194	0.221
CARDINAL HEALTH - OR	ORE-90703	4	14	2	2	-	-	-	-	-	-	-	-	-	22	18	1.161	0.065
CARDINAL HEALTH - OR	ORE-90914	-	6	2	-	-	-	-	-	-	-	-	-	-	8	8	0.280	0.035
CARDINAL HEALTH - PA	10239-02-002	1	16	6	3	-	-	-	-	-	-	-	-	-	26	25	2.717	0.109
CARDINAL HEALTH - PA	10239-03-004	1	-	-	-	-	-	-	-	1	-	-	-	-	1	1	3.925	**3.925**
CARDINAL HEALTH - PA	10239-03-006	1	3	2	1	2	-	1	-	-	-	-	-	-	10	9	2.742	0.305
CARDINAL HEALTH - PA	AC50-58968	-	-	3	-	-	-	-	-	-	-	-	-	-	3	3	0.548	0.183
CARDINAL HEALTH - PA	PA-0384	6	33	4	3	2	-	-	-	-	-	-	-	-	48	42	3.754	0.089
CARDINAL HEALTH - PA	PA-0385	13	43	14	6	5	-	-	-	-	-	-	-	-	**81**	68	8.139	0.120
CARDINAL HEALTH - PA	PA-0415	1	24	3	-	-	-	-	-	-	-	-	-	-	28	27	1.168	0.043
CARDINAL HEALTH - PA	PA-0460	3	13	2	-	-	-	-	-	-	-	-	-	-	18	15	0.555	0.037
CARDINAL HEALTH - PA	PA-0643	5	13	-	-	-	-	-	-	-	-	-	-	-	18	13	0.429	0.033
CARDINAL HEALTH - PA	PA-0680	6	17	3	-	-	-	-	-	-	-	-	-	-	26	20	0.853	0.043
CARDINAL HEALTH - PA	PA-0851	-	-	-	-	-	-	1	-	-	-	-	-	-	1	1	1.779	1.779
CARDINAL HEALTH - PA	PA-0892	6	15	2	-	-	-	-	-	-	-	-	-	-	23	17	0.747	0.044
CARDINAL HEALTH - PA	PA-0893	9	-	-	1	-	-	-	-	-	-	-	-	-	10	1	0.276	0.276
CARDINAL HEALTH - RI	3B-114-01	3	31	3	-	-	-	-	-	-	-	-	-	-	37	34	1.133	0.033
CARDINAL HEALTH - SC	448	4	19	-	-	-	-	-	-	-	-	-	-	-	23	19	0.514	0.027
CARDINAL HEALTH - SD	47111	1	10	3	-	-	-	-	-	-	-	-	-	-	14	13	1.056	0.081
CARDINAL HEALTH - TN	R-19199-B15	5	28	2	-	-	-	-	-	-	-	-	-	-	35	30	1.083	0.036
CARDINAL HEALTH - TN	R-33111-I14	1	16	1	-	-	-	-	-	-	-	-	-	-	18	17	0.481	0.028
CARDINAL HEALTH - TN	R47080-L14	7	19	4	-	-	-	-	-	-	-	-	-	-	30	23	0.940	0.041
CARDINAL HEALTH - TN	R-57025-C15	3	3	2	-	-	-	-	-	-	-	-	-	-	8	5	0.287	0.057
CARDINAL HEALTH - TN	R-71029-B11	2	6	-	-	-	-	-	-	-	-	-	-	-	8	6	0.199	0.033
CARDINAL HEALTH - TN	R-79174-D17	3	12	4	4	-	-	-	-	-	-	-	-	-	23	20	2.134	0.107

APPENDIX A
Table A - Annual TEDE for Agreement State Licensees
2009 (continued)

MANUFACTURING AND DISTRIBUTION – NUCLEAR PHARMACIES – 02500

PROGRAM CODE - LICENSEE NAME	LICENSE#	No Meas. Exposure	Meas. <0.10	0.10-0.25	0.25-0.50	0.50-0.75	0.75-1.00	1.00-2.00	2.00-3.00	3.00-4.00	4.00-5.00	5.00-6.00	6.00-12.00	>12.0	Total Number Monitored	Number with Meas. Dose	Total Collective TEDE (person-rem)	Average Meas. TEDE (rem)
CARDINAL HEALTH - TX	LO-1911	17	32	9	8	1	-	-	-	-	-	-	-	-	67	50	5.989	0.120
CARDINAL HEALTH - TX	LO-1999	2	10	3	-	1	-	-	-	-	-	-	-	-	16	14	1.505	0.108
CARDINAL HEALTH - TX	LO-2033	10	27	9	6	1	-	-	-	-	-	-	-	-	53	43	4.782	0.111
CARDINAL HEALTH - TX	LO-2048	3	31	4	6	1	-	-	-	-	-	-	-	-	45	42	4.084	0.097
CARDINAL HEALTH - TX	LO-2117	2	14	3	-	-	-	-	-	-	-	-	-	-	19	17	0.760	0.045
CARDINAL HEALTH - TX	LO-2737	5	7	1	-	-	-	-	-	-	-	-	-	-	13	8	0.218	0.027
CARDINAL HEALTH - TX	LO-3398	1	10	-	-	-	-	-	-	-	-	-	-	-	11	10	0.243	0.024
CARDINAL HEALTH - TX	LO-4043	4	5	1	-	-	-	-	-	-	-	-	-	-	10	6	0.297	0.050
CARDINAL HEALTH - TX	LO-4573	2	8	1	-	-	-	-	-	-	-	-	-	-	11	9	0.217	0.024
CARDINAL HEALTH - TX	LO-4781	1	25	1	-	-	-	-	-	-	-	-	-	-	27	26	0.374	0.014
CARDINAL HEALTH - TX	LO-4785	5	10	1	-	-	-	-	-	-	-	-	-	-	16	11	0.427	0.039
CARDINAL HEALTH - TX	LO-5461	-	7	-	-	-	-	-	-	-	-	-	-	-	7	7	0.169	0.024
CARDINAL HEALTH - TX	LO-5536	-	1	1	1	-	2	-	-	-	-	-	-	-	5	5	2.351	0.470
CARDINAL HEALTH - TX	LO-5610	-	-	1	2	-	-	-	-	-	-	-	-	-	3	3	0.960	0.320
CARDINAL HEALTH - TX	R33624	3	10	3	2	1	1	-	1	-	-	-	-	-	21	18	5.355	0.298
CARDINAL HEALTH - UNITED STATES	R-79272-111	1	3	-	1	-	-	-	-	-	-	-	-	-	5	4	0.574	0.144
CARDINAL HEALTH - US	065-0794-11	1	6	3	-	-	-	-	-	-	-	-	-	-	10	9	0.794	0.088
CARDINAL HEALTH - US	CH-US-001	18	32	8	6	-	1	-	1	-	-	-	-	-	66	48	7.629	0.159
CARDINAL HEALTH - UT	UT-1800225	6	18	2	-	-	-	-	-	-	-	-	-	-	26	20	0.893	0.045
CARDINAL HEALTH - VA	760-34-1	6	45	7	1	-	-	-	-	-	-	-	-	-	59	53	2.534	0.048
CARDINAL HEALTH - WA	WN-NP003-1	3	28	5	-	-	-	-	-	-	-	-	-	-	36	33	1.599	0.048
CARDINAL HEALTH - WA	WN-NP004-1	4	21	2	1	-	-	-	-	-	-	-	-	-	28	24	1.231	0.051
CARDINAL HEALTH - WA	WN-NP005-1	2	17	-	-	-	-	-	-	-	-	-	-	-	19	17	0.575	0.034
CARDINAL HEALTH - WA	WN-NP006-1	-	4	2	-	-	-	-	-	-	-	-	-	-	6	6	0.495	0.083
CARDINAL HEALTH - WA	WN-NP011-1	2	4	1	1	1	2	-	-	1	-	-	-	-	12	10	5.871	0.587
CARDINAL HEALTH - WI	025-1123-01	2	20	6	-	-	-	-	-	-	-	-	-	-	28	26	1.290	0.050
CARDINAL HEALTH - WI	087-1312-01	15	14	3	-	-	-	-	-	-	-	-	-	-	32	17	0.810	0.048
CARDINAL HEALTH - WI	141-1306-01	5	12	-	-	-	-	-	-	-	-	-	-	-	17	12	0.241	0.020
GE HEALTHCARE	PA-0515	23	13	7	-	-	-	-	-	-	-	-	-	-	43	20	1.596	0.080
GE HEALTHCARE - ANAHEIM	4810-30	18	9	-	-	-	-	-	-	-	-	-	-	-	27	9	0.266	0.030
GE HEALTHCARE - SACRAMENTO	4809-34	22	7	-	-	-	-	-	-	-	-	-	-	-	29	7	0.186	0.027
GE HEALTHCARE - SAN DIEGO	5796-37	9	10	-	-	-	-	-	-	-	-	-	-	-	19	10	0.331	0.033
GE HEALTHCARE - SAN JOSE	4811-43	4	6	5	2	-	-	-	-	-	-	-	-	-	17	13	1.681	0.129

APPENDIX A

Table A - Annual TEDE for Agreement State Licensees
2009 (continued)

PROGRAM CODE - LICENSEE NAME	LICENSE#	No Meas. Exposure	Meas. <0.10	Number of Individuals with Whole Body Doses in the Ranges (rems)*											Total Number Monitored	Number with Meas. Dose	Total Collective TEDE (person-rem)	Average Meas. TEDE (rem)
				0.10-0.25	0.25-0.50	0.50-0.75	0.75-1.00	1.00-2.00	2.00-3.00	3.00-4.00	4.00-5.00	5.00-6.00	6.00-12.00	>12.0				
MANUFACTURING AND DISTRIBUTION – NUCLEAR PHARMACIES – 02500																		
GE HEALTHCARE - VAN NUYS	5143-19	11	7	1	-	-	-	-	-	-	-	-	-	-	19	8	0.364	0.046
IBA MOLECULAR NORTH AMERICA, INC.	CA-7131-43	4	14	8	2	2	2	3	1	-	-	-	-	-	36	32	11.199	0.350
MALLINCKRODT. INC	859-01	8	3	2	-	-	-	-	-	-	-	-	-	-	13	5	0.500	0.100
MALLINCKRODT. INC	PA-0541	4	3	2	2	-	-	-	-	-	-	-	-	-	11	7	1.080	0.154
MALLINCKRODT. INC.	PA-0842	10	3	2	1	-	-	-	-	-	-	-	-	-	16	6	0.843	0.141
MORAVEK BIOCHEMICALS. INC.	2960-30	2	23	6	-	1	-	1	-	-	-	-	-	-	33	31	4.340	0.140
TRIAD ISOTOPES. INC.	PA-0479	2	12	4	2	-	-	-	-	-	-	-	-	-	20	18	1.931	0.107
TRIAD ISOTOPES. INC.	PA-0637	10	4	-	-	-	-	-	-	-	-	-	-	-	14	4	0.181	0.045
Total	171	818	2,472	492	185	64	36	27	9	2	-	-	-	-	4,105	3,287	340.924	0.104
WELL LOGGING BYPRODUCT AND/OR SNM SEALED SOURCES ONLY - 03111																		
ISOTOPE PRODUCTS LABS	1509-19	3	22	15	3	6	8	11	-	-	-	-	-	-	68	65	30.503	0.469
Total	1	3	22	15	3	6	8	11	-	-	-	-	-	-	68	65	30.503	0.469
MANUFACTURING AND DISTRIBUTION – OTHER - 03214																		
HOCHIKI AMERICA CORPORATION	2090-30	18	-	-	-	-	-	-	-	-	-	-	-	-	18	-	-	-
J. L. SHEPHERD AND ASSOCIATES	CA 1777-19	14	7	3	2	-	-	-	-	-	-	-	-	-	26	12	1.365	0.114
Total	2	32	7	3	2	-	-	-	-	-	-	-	-	-	44	12	1.365	0.114
INSTRUMENT CALIBRATION SERVICE ONLY – SOURCE > 100 CURIES - 03222																		
EXELON POWERLABS	PA-1017	34	-	-	-	-	-	-	-	-	-	-	-	-	34	-	-	-
Total	1	34	-	-	-	-	-	-	-	-	-	-	-	-	34	-	-	-
OTHER SERVICES - 03225																		
HYSPAN PRECISION PRODUCTS, INC.	7010-37	-	1	-	-	-	-	1	-	-	-	-	-	-	2	2	0.414	0.207
Total	1	-	1	-	-	-	-	1	-	-	-	-	-	-	2	2	0.414	0.207
WASTE DISPOSAL SERVICE PROCESSING AND/OR REPACKAGING - 03234																		
ENVIRONMENTAL MANAGEMENT & CONTROLS	3546-50	1	-	-	-	1	-	1	-	-	-	-	-	-	3	2	1.712	0.856
Total	1	1	-	-	-	1	-	1	-	-	-	-	-	-	3	2	1.712	0.856
INDUSTRIAL RADIOGRAPHY – FIXED LOCATION – 03310																		
SONGS - RADIOGRAPHY GROUP	CA-5244-30	8	-	1	-	-	-	-	-	-	-	-	-	-	9	1	0.126	0.126
Total	1	8	-	1	-	-	-	-	-	-	-	-	-	-	9	1	0.126	0.126

APPENDIX A
Table A - Annual TEDE for Agreement State Licensees
2009 (continued)

PROGRAM CODE - LICENSEE NAME	LICENSE#	Number of Individuals with Whole Body Doses in the Ranges (rems)*													Total Number Monitored	Number with Meas. Dose	Total Collective TEDE (person-rem)	Average Meas. TEDE (rem)
		No Meas. Exposure	Meas. <0.10	0.10-0.25	0.25-0.50	0.50-0.75	0.75-1.00	1.00-2.00	2.00-3.00	3.00-4.00	4.00-5.00	5.00-6.00	6.00-12.00	>12.0				
INDUSTRIAL RADIOGRAPHY – TEMPORARY JOB SITE – 03320																		
ADVEX CORPORATION	650-254-1	2	3	1	1	-	-	-	-	-	-	-	-	-	7	5	0.522	0.104
ALARON CORPORATION	PA-0678	8	69	8	13	9	4	9	3	-	-	-	-	-	123	115	36.528	0.318
ALPHA-OMEGA SERVICES, INC.	3925-19	7	10	3	-	-	-	-	-	-	-	-	-	-	20	13	0.588	0.045
AUTOMATION AND CONTROL TECHNOLOGY	03214250031	6	11	2	-	-	-	-	-	-	-	-	-	-	19	13	0.905	0.070
BAKER INSPECTION GROUP	OH-0332077	4	1	1	1	-	-	-	-	-	-	-	-	-	7	3	0.497	0.166
BRAUN INTERTEC CORPORATION	1082-103-27	6	10	1	3	1	4	10	2	1	-	-	-	-	38	32	29.913	0.935
CERTIFIED TESTING LABORATORIES, INC.	507161	4	8	1	1	1	-	-	-	-	-	-	-	-	15	11	1.261	0.115
CERTIFIED TESTING LABORATORIES, INC.	NYS C1920	2	5	4	1	-	-	-	-	-	-	-	-	-	12	10	1.333	0.133
CONSTRUCTION MATERIALS TESTING, INC.	0799-07	4	3	-	-	-	-	-	-	-	-	-	-	-	7	3	0.073	0.024
COOPERHEAT-MQS INSPECTION	388-01	5	5	3	10	5	5	4	-	-	-	-	-	-	37	32	17.403	0.544
CTL - ASTROTECH DIVISION	PA-0430	12	7	2	-	-	-	-	-	-	-	-	-	-	21	9	0.455	0.051
EDGE INSPECTION GROUP, INC	CA-7214-48	-	7	1	6	3	2	1	-	-	-	-	-	-	20	20	7.036	0.352
EG&G TECHNICAL SERVICES, INC	EGG-4077-1	18	3	2	-	-	-	-	-	-	-	-	-	-	23	5	0.395	0.079
EWER TESTING & INSPECTION, INC.	ND-33-32613	1	1	-	2	2	2	1	-	-	-	-	-	-	9	8	5.349	0.669
INTERNATIONAL INSPECTION, INC	5046-19	4	4	1	-	1	1	-	-	-	-	-	-	-	11	7	1.081	0.154
MIDWEST INDUSTRIAL X-RAY, INC	ND-33-27427	-	-	-	-	3	3	10	2	-	-	-	-	-	19	19	23.312	1.227
MISTRAS GROUP, INC.	4886-48	4	16	4	14	14	11	13	-	-	-	-	-	-	76	72	41.389	0.575
NDT LABORATORIES, INC.	1651-43	3	2	1	-	-	-	-	-	-	-	-	-	-	6	3	0.283	0.094
NORTHROP GRUMMAN	0043 43	10	-	-	-	-	-	-	-	-	-	-	-	-	10	-	-	-
POLE BROTHERS IMAGING COMPANY	069-451-1	-	1	1	-	1	-	-	-	-	-	-	-	-	3	3	0.767	0.256
PRECISION CUSTOM COMPONENTS	PA-1042	17	2	-	-	-	-	-	-	-	-	-	-	-	19	2	0.002	0.001
QUALITY INSPECTION SERVICES, INC	3043-1	-	8	-	-	2	1	-	-	-	-	-	-	-	11	11	2.235	0.203
QUALITY INSPECTION SERVICES, INC	3043-2	-	1	2	-	-	-	-	-	-	-	-	-	-	3	3	0.445	0.148

APPENDIX A

Table A - Annual TEDE for Agreement State Licensees
2009 (continued)

PROGRAM CODE - LICENSEE NAME	LICENSE#	No Meas. Exposure	Meas. <0.10	0.10-0.25	0.25-0.50	0.50-0.75	0.75-1.00	1.00-2.00	2.00-3.00	3.00-4.00	4.00-5.00	5.00-6.00	6.00-12.00	>12.0	Total Number Monitored	Number with Meas. Dose	Total Collective TEDE (person-rem)	Average Meas. TEDE (rem)
INDUSTRIAL RADIOGRAPHY – TEMPORARY JOB SITE – 03320																		
QUALITY INSPECTION SERVICES, INC.	C3267	1	3	-	-	1	-	1	-	-	-	-	-	-	6	5	1.750	0.350
QUALITY INSPECTION SERVICES, INC.	LO-6219	1	1	2	4	1	1	2	-	-	-	-	-	-	12	11	5.619	0.511
QUALITY INSPECTION SERVICES, INC.	NYS C2514	1	10	1	6	3	2	5	-	-	-	-	-	-	28	27	14.437	0.535
QUALITY INSPECTION SERVICES, INC.	NYS C2700	-	2	2	6	1	1	1	-	-	-	-	-	-	13	13	5.027	0.387
QUALITY INSPECTION SERVICES, INC.	NYS DOH C2505	-	-	3	-	-	1	-	-	-	-	-	-	-	4	4	0.962	0.241
QUALITY INSPECTION SERVICES, INC.	PA-1350	-	4	2	1	-	-	-	-	-	-	-	-	-	7	7	1.130	0.161
SPX HEAT TRANSFER, INC.	OK-13735	-	3	-	-	-	-	-	-	-	-	-	-	-	3	3	0.068	0.023
SUBSURFACE IMAGING, INC.	7449-19	2	7	2	5	1	-	-	-	-	-	-	-	-	17	15	3.581	0.239
TC INSPECTION, INC.	5299-07	32	17	13	14	10	7	8	3	1	-	-	-	-	105	73	41.805	0.573
TESTING ENGINEERS, INC.	1898-01	1	2	1	-	1	-	-	-	-	-	-	-	-	5	4	0.846	0.212
THERMO GAMMA METRICS	3775-37	58	11	1	-	-	-	-	-	-	-	-	-	-	70	12	0.530	0.044
THREE RIVERS GAMMA SERVICES	PA-1165	-	-	-	-	-	-	2	-	-	-	-	-	-	2	2	2.850	1.425
VALLEY INDUSTRIAL X-RAY & INSPECTION	CA-4182-15	16	33	20	11	18	12	28	7	-	-	-	-	-	145	129	87.723	0.680
WESTERN INDUSTRIAL X-RAY	4424-48	1	2	1	4	3	-	1	-	-	-	-	-	-	12	11	4.514	0.410
WESTERN X-RAY CORPORATION	5324-56	5	2	3	-	1	-	1	-	-	-	-	-	-	12	7	2.235	0.319
Total	**38**	235	274	89	105	80	58	97	17	2	-	-	-	-	957	722	344.849	0.478
IRRADIATORS - OTHER - LESS THAN 10000 CURIES - 03511																		
INDUSTRIAL NUCLEAR CO. INC.	2229-01	6	-	1	1	1	-	1	-	-	-	-	-	-	10	4	2.113	0.528
Total	**1**	6	-	1	1	1	-	1	-	-	-	-	-	-	10	4	2.113	0.528
NO PROGRAM CODE																		
BUFFALO GASTROENTEROLOGY ASSOCIATES, LLP	1363121	-	1	-	-	-	-	-	-	-	-	-	-	-	1	1	0.014	0.014
Total	**1**	-	1	-	-	-	-	-	-	-	-	-	-	-	1	1	0.014	0.014

Number of Individuals with Whole Body Doses in the Ranges (rems)*

APPENDIX A

Table A - Annual TEDE for Agreement State Licensees 2010

PROGRAM CODE - LICENSEE NAME	LICENSE#	No Meas. Exposure	Meas. <0.10	0.10-0.25	0.25-0.50	0.50-0.75	0.75-1.00	1.00-2.00	2.00-3.00	3.00-4.00	4.00-5.00	5.00-6.00	6.00-12.00	>12.0	Total Number Monitored	Number with Meas. Dose	Total Collective TEDE (person-rem)	Average Meas. TEDE (rem)
VETERINARY NON-HUMAN - 02400																		
PHOENIX CENTRAL LAB FOR VET	WNL-021C	4	-	-	-	-	-	-	-	-	-	-	-	-	4	-	-	-
Total	1	4	-	-	-	-	-	-	-	-	-	-	-	-	4	-	-	-
MANUFACTURING AND DISTRIBUTION – NUCLEAR PHARMACIES – 02500																		
CARDINAL HEALTH - AK	0707RA900-912	-	5	-	-	-	-	-	-	-	-	-	-	-	5	5	0.125	0.025
CARDINAL HEALTH - CT	00163	4	40	8	3	-	-	-	-	-	-	-	-	-	55	51	3.504	0.069
CARDINAL HEALTH - IN	R0358-45	19	23	-	-	-	-	-	-	-	-	-	-	-	42	23	0.482	0.021
CARDINAL HEALTH - MI	1549-4	16	27	4	1	-	-	-	-	-	-	-	-	-	47	31	1.123	0.036
CARDINAL HEALTH - MO	2840	11	35	7	1	3	1	-	-	-	-	-	-	-	58	47	5.496	0.117
CARDINAL HEALTH - PA	PA-0892	15	11	1	-	-	-	-	-	-	-	-	-	-	27	12	0.539	0.045
CARDINAL HEALTH - PA	PA-0893	3	1	1	-	-	-	-	-	-	-	-	-	-	5	2	0.197	0.099
CARDINAL HEALTH - SD	47111	-	11	3	-	-	-	-	-	-	-	-	-	-	14	14	1.222	0.087
GE HEALTHCARE	PA-0515	17	21	7	7	-	-	-	-	-	-	-	-	-	52	35	4.249	0.121
GE HEALTHCARE - ANAHEIM	4810-30	11	7	2	-	-	-	-	-	-	-	-	-	-	20	9	0.367	0.041
GE HEALTHCARE - SACRAMENTO	4809-34	20	9	1	-	-	-	-	-	-	-	-	-	-	30	10	0.459	0.046
GE HEALTHCARE - SAN DIEGO	5796-37	13	4	-	-	-	-	-	-	-	-	-	-	-	17	4	0.155	0.039
GE HEALTHCARE - SAN JOSE	4811-43	9	7	4	1	-	-	-	-	-	-	-	-	-	21	12	1.225	0.102
GE HEALTHCARE - VAN NUYS	5143-19	11	6	1	-	-	-	-	-	-	-	-	-	-	18	7	0.280	0.040
IBA MOLECULAR NORTH AMERICA, INC	CA-7131-43	8	6	13	5	1	2	1	1	1	-	-	-	-	38	30	13.684	0.456
MALLINCKRODT, INC.	PA-0842	4	2	2	1	-	-	-	-	-	-	-	-	-	9	5	0.636	0.127
PANTEX DIVISION OF BIOANALYSIS INC	CA-2402-19	4	2	-	-	-	-	-	-	-	-	-	-	-	6	2	0.024	0.012
TRIAD ISOTOPES INC	PA-0418	10	4	-	-	-	-	-	-	-	-	-	-	-	14	4	0.134	0.034
TRIAD ISOTOPES INC	PA-0479	5	10	8	2	-	-	-	-	-	-	-	-	-	25	20	2.382	0.119
Total	19	180	231	62	20	4	3	1	1	1	-	-	-	-	503	323	36.283	0.112
WELL LOGGING BYPRODUCT AND/OR SNM SEALED SOURCES ONLY - 03111																		
ISOTOPE PRODUCTS LABS	1509-19	12	18	10	5	6	8	9	-	-	-	-	-	-	68	56	27.050	0.483
Total	1	12	18	10	5	6	8	9	-	-	-	-	-	-	68	56	27.050	0.483
MEASURING SYSTEMS PORTABLE GAUGES - 03121																		
PROFESSIONAL SERVICE INDUSTRIES INC	PA-0281	4	3	3	3	1	1	-	-	-	-	-	-	-	15	11	3.900	0.355
Total	1	4	3	3	3	1	1	-	-	-	-	-	-	-	15	11	3.900	0.355

Number of Individuals with Whole Body Doses in the Ranges (rems)

APPENDIX A

Table A - Annual TEDE for Agreement State Licensees 2010 (continued)

PROGRAM CODE - LICENSEE NAME	LICENSE#	No Meas. Exposure	Number of Individuals with Whole Body Doses in the Ranges (rems)[a]												Total Number Monitored	Number with Meas. Dose	Total Collective TEDE (person-rem)	Average Meas. TEDE (rem)
			Meas. <0.10	0.10-0.25	0.25-0.50	0.50-0.75	0.75-1.00	1.00-2.00	2.00-3.00	3.00-4.00	4.00-5.00	5.00-6.00	6.00-12.00	>12.0				
MANUFACTURING AND DISTRIBUTION – OTHER - 03214																		
HOCHIKI AMERICA CORPORATION	2090-30	20	-	-	-	-	-	-	-	-	-	-	-	-	20	-	-	-
J. L. SHEPHERD AND ASSOCIATES	CA 1777-19	17	3	-	4	1	1	1	-	-	-	-	-	-	27	10	4.709	0.471
Total	2	37	3	-	4	1	1	1	-	-	-	-	-	-	47	10	4.709	0.471
INSTRUMENT CALIBRATION SERVICE ONLY – SOURCE > 100 CURIES - 03222																		
EXELON POWERLABS	PA-1017	35	-	-	-	-	-	-	-	-	-	-	-	-	35	-	-	-
Total	1	35	-	-	-	-	-	-	-	-	-	-	-	-	35	-	-	-
OTHER SERVICES - 03225																		
HYSPAN PRECISION PRODUCTS, INC.	7010-37	-	1	1	-	-	-	-	-	-	-	-	-	-	2	2	0.247	0.124
Total	1	-	1	1	-	-	-	-	-	-	-	-	-	-	2	2	0.247	0.124
WASTE DISPOSAL SERVICE PROCESSING AND/OR REPACKAGING - 03234																		
ENVIRONMENTAL MANAGEMENT & CONTROLS	3546-50	-	1	-	-	1	-	-	1	-	-	-	-	-	3	3	2.681	0.894
Total	1	-	1	-	-	1	-	-	1	-	-	-	-	-	3	3	2.681	0.894
INDUSTRIAL RADIOGRAPHY – FIXED LOCATION – 03310																		
SONGS - RADIOGRAPHY GROUP	CA-5244-30	4	6	-	-	-	-	-	-	-	-	-	-	-	10	6	0.083	0.014
Total	1	4	6	-	-	-	-	-	-	-	-	-	-	-	10	6	0.083	0.014
INDUSTRIAL RADIOGRAPHY – TEMPORARY JOB SITE – 03320																		
ADVEX CORPORATION	650-254-1	-	2	1	1	-	-	-	-	-	-	-	-	-	4	4	0.606	0.152
ALARON CORPORATION	PA-0678	9	74	12	1	-	-	-	-	-	-	-	-	-	96	87	3.890	0.045
ALPHA-OMEGA SERVICES, INC.	3925-19	13	6	4	2	-	-	-	-	-	-	-	-	-	25	12	1.566	0.131
AUTOMATION AND CONTROL TECHNOLOGY	03214250001	2	11	4	-	-	-	-	-	-	-	-	-	-	17	15	0.740	0.049
BAKER INSPECTION GROUP	OH-0332077	1	7	1	1	-	1	1	-	-	-	-	-	-	12	11	3.167	0.288
BRAUN INTERTEC CORPORATION	1082-103-27	3	12	3	8	5	4	10	1	-	-	-	-	-	46	43	27.919	0.649
CODE INSPECTION & TESTING CO	PA-1415	1	-	2	1	2	-	-	-	-	-	-	-	-	6	5	1.718	0.344
CONSTRUCTION MATERIALS TESTING, INC	0799-07	3	3	-	1	-	-	-	-	-	-	-	-	-	7	4	0.316	0.079
CTL - ASTROTECH DIVISION	PA-0430	1	12	1	1	-	-	-	-	-	-	-	-	-	15	14	0.739	0.053
DOMINION NDT SERVICES, INC.	041-527-1	-	2	2	2	-	-	-	-	-	-	-	-	-	6	6	1.069	0.178
EDGE INSPECTION GROUP INC	CA-7214-48	6	8	2	5	1	-	1	-	-	-	-	-	-	23	17	4.271	0.251
EG&G TECHNICAL SERVICES, INC	EGG-4077-1	22	2	-	-	-	-	-	-	-	-	-	-	-	24	2	0.100	0.050
ENGINEERING & INSPECTIONS, INC	PA-1080	-	-	4	1	1	2	4	-	-	-	-	-	-	12	12	9.861	0.822
INSPECTION TESTING IND RADIOGRAPHY OPS	CA-5903-37	16	6	-	-	-	-	-	-	-	-	-	-	-	22	6	0.093	0.016
INTERNATIONAL ENERGY SERVICES, CO	CA-6571-19	26	17	10	17	3	4	1	-	-	-	-	-	-	78	52	15.851	0.305
INTERNATIONAL INSPECTION, INC	5046-19	7	3	1	-	1	-	-	-	-	-	-	-	-	12	5	1.106	0.221
MISTRAS GROUP INC	CA-4886-48	8	27	9	17	9	8	6	1	-	-	-	-	-	85	77	31.241	0.406

APPENDIX A

Table A - Annual TEDE for Agreement State Licensees

2010 (continued)

PROGRAM CODE - LICENSEE NAME	LICENSE#	No Meas. Exposure	Number of Individuals with Whole Body Doses in the Ranges (rems)*												Total Number Monitored	Number with Meas. Dose	Total Collective TEDE (person-rem)	Average Meas. TEDE (rem)
			Meas. <0.10	0.10-0.25	0.25-0.50	0.50-0.75	0.75-1.00	1.00-2.00	2.00-3.00	3.00-4.00	4.00-5.00	5.00-6.00	6.00-12.00	>12.0				
INDUSTRIAL RADIOGRAPHY – TEMPORARY JOB SITE – 03320																		
MISTRAS GROUP, INC.	PA-1138	4	16	9	7	4	4	9	-	-	-	-	-	-	53	49	23.793	0.486
NAASCO	CA-0684-37		7	-	-	-	-	-	-	-	-	-	-	-	7	7	0.084	0.012
NDT LABORATORIES, INC.	1651-43	4	-	2	1	-	-	-	-	-	-	-	-	-	7	3	0.546	0.182
PRECISION CUSTOM COMPONENTS	PA-1042	10	9	-	-	-	-	-	-	-	-	-	-	-	19	9	0.021	0.002
QUALITY INSPECTION SERVICES, INC	4196-1		3	2	-	-	-	-	-	-	-	-	-	-	5	5	0.452	0.090
QUALITY INSPECTION SERVICES, INC	4196-2		2	1	1	1	-	-	-	-	-	-	-	-	5	5	1.755	0.351
QUALITY INSPECTION SERVICES, INC	C3267		5	-	1	1	-	-	-	-	-	-	-	-	7	7	1.096	0.157
QUALITY INSPECTION SERVICES, INC	IR-074-01		6	1	4	-	-	-	-	-	-	-	-	-	11	11	1.864	0.169
QUALITY INSPECTION SERVICES, INC	LO-6219	1	3	3	5	-	-	1	-	-	-	-	-	-	13	12	3.607	0.301
QUALITY INSPECTION SERVICES, INC	NYS C2514	1	7	2	2	7	1	5	1	-	-	-	-	-	26	25	15.537	0.621
QUALITY INSPECTION SERVICES, INC	NYS C2700	1	2	3	3	-	-	1	-	-	-	-	-	-	10	9	3.024	0.336
QUALITY INSPECTION SERVICES, INC	NYS DOH C2505		5	1	1	6	-	1	-	-	-	-	-	-	14	14	5.351	0.382
QUALITY INSPECTION SERVICES, INC	PA-1350		2	2	2	-	-	-	-	-	-	-	-	-	6	6	0.885	0.148
SPX HEAT TRANSFER, INC.	OK-13735		2	-	-	-	-	-	-	-	-	-	-	-	2	2	0.121	0.061
STORK MATERIALS TESTING & INSPECTION	CA-1880-1E		-	1	3	-	2	-	-	-	-	-	-	-	6	6	2.872	0.479
SUBSURFACE IMAGING, INC.	7449-19	2	3	3	4	1	2	-	-	-	-	-	-	-	15	13	4.202	0.323
TC INSPECTION, INC.	5299-07	32	24	7	21	7	5	3	-	-	-	-	-	-	100	68	26.004	0.382
TESTING ENGINEERS, INC.	1898-01		1	2	3	-	-	-	-	-	-	-	-	-	6	6	1.335	0.223
THREE RIVERS GAMMA SERVICES	PA-1165		-	-	-	-	-	2	-	-	-	-	-	-	2	2	3.117	1.559
VALLEY INDUSTRIAL X-RAY & INSPECTION	CA-4182-1E	41	35	24	24	20	20	30	13	1	-	-	-	-	207	166	114.479	0.690
VALLEY INSPECTION SERVICE, INC.	PA-1186		1	1	2	-	1	1	1	1	-	-	-	-	7	7	8.723	1.246
WESTERN INDUSTRIAL X-RAY	4424-48		2	1	-	1	4	1	-	-	-	-	-	-	9	9	5.387	0.599
WESTEX COMPANY	5324-56	3	2	1	-	2	-	-	-	-	-	-	-	-	8	5	1.464	0.293
Total	**40**	217	329	121	141	72	59	77	17	2	-	-	-	-	1,035	818	329.972	0.403

NRC FORM 335 (12-2010) NRCMD 3.7	U.S. NUCLEAR REGULATORY COMMISSION	1. REPORT NUMBER (Assigned by NRC, Add Vol., Supp., Rev., and Addendum Numbers, if any.)
	BIBLIOGRAPHIC DATA SHEET *(See Instructions on the reverse)*	NUREG-2118, Volume 1

2. TITLE AND SUBTITLE	3. DATE REPORT PUBLISHED	
Occupational Radiation Exposure at Agreement State-Licensed Materials Facilities, 1997-2010	MONTH July	YEAR 2012
	4. FIN OR GRANT NUMBER	

5. AUTHOR(S)	6. TYPE OF REPORT
D.E. Lewis * D.A. Hagemeyer *Y.U. McCormick	Bi-Annual
	7. PERIOD COVERED (Inclusive Dates) Calendars Years 1997-2010

8. PERFORMING ORGANIZATION - NAME AND ADDRESS (If NRC, provide Division, Office or Region, U. S. Nuclear Regulatory Commission, and mailing address; if contractor, provide name and mailing address.)

Division of Systems Analysis
Office of Nuclear Regulatory Research
US Nuclear Regulatory Commission
Washington, DC 20555-0001

*Oak Ridge Associated Universities (ORAU)
1299 Bethel Valley Road, SC-200, MS-21
Oak Ridge, TN 37830

9. SPONSORING ORGANIZATION - NAME AND ADDRESS (If NRC, type "Same as above"; if contractor, provide NRC Division, Office or Region, U. S. Nuclear Regulatory Commission, and mailing address.)

Same as 8 above.

10. SUPPLEMENTARY NOTES

11. ABSTRACT (200 words or less)

This report summarizes the occupational radiation exposure data maintained in the U.S. Nuclear Regulatory Commission's (NRC's) Radiation Exposure Information and Reporting System (REIRS). The bulk of the information contained in this report was compiled from the 1997 through 2010 annual reports submitted by nine categories of Agreement State licensees. Agreement State licensees are not subject to the reporting requirements in 10 CFR 20.2206. However, these licensees voluntarily submitted this data as part of the agency's information request regarding potential changes to 10 CFR Part 20. The annual reports submitted by these licensees consist of radiation exposure records for each monitored individual.

Annual reports were received from a total of 312 Agreement State licensees. Compilations of the reports submitted by the 312 Agreement State licensees indicated that 40,622 individuals were monitored, 31,704 of whom received a measurable dose. The collective dose incurred by these individuals was 5,908 person-rem and the average measureable dose was 0.19 rem.

12. KEY WORDS/DESCRIPTORS (List words or phrases that will assist researchers in locating the report.)	13. AVAILABILITY STATEMENT unlimited
occupational exposure Agreement State	14. SECURITY CLASSIFICATION
	(This Page) unclassified
	(This Report) unclassified
	15. NUMBER OF PAGES
	16. PRICE

NRC FORM 335 (12-2010)

Printed
on recycled
paper

Federal Recycling Program

UNITED STATES
NUCLEAR REGULATORY COMMISSION
WASHINGTON, DC 20555-0001

OFFICIAL BUSINESS

NUREG-2118, Vol. 1

Occupational Radiation Exposure at Agreement State-Licensed
Materials Facilities, 1997-2010

July 2012

www.ingramcontent.com/pod-product-compliance
Lightning Source LLC
Chambersburg PA
CBHW080304180526
45167CB00006B/2667